OPPORTUNITIES TO ADDRESS CLINICAL RESEARCH WORKFORCE DIVERSITY NEEDS FOR 2010

Committee on Opportunities to Address
Clinical Research Workforce Diversity Needs for 2010

Committee on Women in Science and Engineering
Policy and Global Affairs

Board on Health Sciences Policy
Institute of Medicine

Jong-on Hahm and Alexander Ommaya, Editors

NATIONAL RESEARCH COUNCIL *AND*
INSTITUTE OF MEDICINE
OF THE NATIONAL ACADEMIES

THE NATIONAL ACADEMIES PRESS
Washington, D.C.
www.nap.edu

THE NATIONAL ACADEMIES PRESS 500 Fifth Street, N.W. Washington, DC 20001

NOTICE: The project that is the subject of this report was approved by the Governing Board of the National Research Council, whose members are drawn from the councils of the National Academy of Sciences, the National Academy of Engineering, and the Institute of Medicine. The members of the committee responsible for the report were chosen for their special competences and with regard for appropriate balance.

This project was supported by the National Institutes of Health, Grant No. N01-OD-4-2139, Task Order #142, and the National Academy of Sciences. Any opinions, findings, conclusions, or recommendations expressed in this publication are those of the author(s) and do not necessarily reflect the views of the organizations or agencies that provided support for the project.

International Standard Book Number 0-309-09248-5

Additional copies of this report are available from the National Academies Press, 500 Fifth Street, N.W., Lockbox 285, Washington, DC 20055; (800) 624-6242 or (202) 334-3313 (in the Washington metropolitan area); Internet, http://www.nap.edu.

Copyright 2006 by the National Academy of Sciences. All rights reserved.

Printed in the United States of America

THE NATIONAL ACADEMIES
Advisers to the Nation on Science, Engineering, and Medicine

The **National Academy of Sciences** is a private, nonprofit, self-perpetuating society of distinguished scholars engaged in scientific and engineering research, dedicated to the furtherance of science and technology and to their use for the general welfare. Upon the authority of the charter granted to it by the Congress in 1863, the Academy has a mandate that requires it to advise the federal government on scientific and technical matters. Dr. Ralph J. Cicerone is president of the National Academy of Sciences.

The **National Academy of Engineering** was established in 1964, under the charter of the National Academy of Sciences, as a parallel organization of outstanding engineers. It is autonomous in its administration and in the selection of its members, sharing with the National Academy of Sciences the responsibility for advising the federal government. The National Academy of Engineering also sponsors engineering programs aimed at meeting national needs, encourages education and research, and recognizes the superior achievements of engineers. Dr. Wm. A. Wulf is president of the National Academy of Engineering.

The **Institute of Medicine** was established in 1970 by the National Academy of Sciences to secure the services of eminent members of appropriate professions in the examination of policy matters pertaining to the health of the public. The Institute acts under the responsibility given to the National Academy of Sciences by its congressional charter to be an adviser to the federal government and, upon its own initiative, to identify issues of medical care, research, and education. Dr. Harvey V. Fineberg is president of the Institute of Medicine.

The **National Research Council** was organized by the National Academy of Sciences in 1916 to associate the broad community of science and technology with the Academy's purposes of furthering knowledge and advising the federal government. Functioning in accordance with general policies determined by the Academy, the Council has become the principal operating agency of both the National Academy of Sciences and the National Academy of Engineering in providing services to the government, the public, and the scientific and engineering communities. The Council is administered jointly by both Academies and the Institute of Medicine. Dr. Ralph J. Cicerone and Dr. Wm. A. Wulf are chair and vice chair, respectively, of the National Research Council.

www.national-academies.org

COMMITTEE ON OPPORTUNITIES TO ADDRESS CLINICAL RESEARCH WORKFORCE DIVERSITY NEEDS FOR 2010

E. Albert Reece, M.D., *Chair*, Vice Chancellor and Dean, University of Arkansas College of Medicine
Rick Martinez, M.D., Director of Medical Affairs, Johnson and Johnson
Nancy E. Reame, Ph.D., Mary Dickey Lindsay Professor of Nursing and Director, DNSc Program, Columbia University
Sally Shaywitz, M.D., Co-director, Yale Center for the Study of Learning and Attention, Yale University School of Medicine
Nancy Sung, Ph.D., Senior Program Officer, Burroughs Wellcome Fund

NRC Staff

Jong-on Hahm, Ph.D., Study Director
Elizabeth Briggs, Senior Program Associate

IOM Staff

Alex Ommaya, Sc.D., Senior Program Officer
Michelle Lyons, M.S., Research Associate (until December 2004)
Amy Haas, Senior Program Assistant

COMMITTEE ON WOMEN IN SCIENCE AND ENGINEERING

Lilian Wu, *Chair*, Director of University Relations, International Business Machines
Lotte Bailyn, T. Wilson Professor of Management, Sloan School of Management, Massachusetts Institute of Technology
Ilene Busch-Vishniac, Professor, Mechanical Engineering, The Johns Hopkins University
Ralph J. Cicerone, Former Chancellor, University of California, Irvine (until January 2005)
Allan Fisher, President and CEO, iCarnegie, Inc.
Sally Shaywitz, Co-director, Yale Center for the Study of Learning and Attention, Yale University School of Medicine
Julia Weertman, Professor Emerita, Department of Material Science and Engineering, Northwestern University

Staff

Jong-on Hahm, Director (until October 14, 2005)
Peter Henderson, Acting Director (from October 15, 2005)
Charlotte Kuh, Deputy Executive Director, Policy and Global Affairs
John Sislin, Program Officer
Elizabeth Briggs Huthnance, Senior Program Associate
Amaliya Jurta, Senior Program Assistant (through July 2002)

BOARD ON HEALTH SCIENCES POLICY

Fred H. Gage, *Chair*, The Salk Institute for Biological Studies, La Jolla, California
Gail H. Cassell, Eli Lilly and Company, Indianapolis, Indiana
James F. Childress, University of Virginia, Charlottesville
Ellen Wright Clayton, Vanderbilt University Medical School, Nashville, Tennessee
David R. Cox, Perlegen Sciences, Mountain View, California
Lynn R. Goldman, Johns Hopkins Bloomberg School of Public Health, Baltimore, Maryland
Bernard D. Goldstein, University of Pittsburgh, Pittsburgh, Pennsylvania
Martha N. Hill, Johns Hopkins University School of Nursing, Baltimore, Maryland
Alan Leshner, American Association for the Advancement of Science, Washington, D.C.
Daniel Masys, Vanderbilt University Medical Center, Nashville, Tennessee
Jonathan D. Moreno, University of Virginia, Charlottesville
E. Albert Reece, University of Arkansas, Little Rock
Myrl Weinberg, National Health Council, Washington, D.C.
Michael J. Welch, Washington University School of Medicine, St. Louis, Missouri
Owen N. Witte, University of California, Los Angeles
Mary Woolley, Research!America, Alexandria, Virginia

IOM Staff

Andrew M. Pope, Director
Amy Haas, Board Assistant
David Codrea, Financial Associate

Preface

Increasing diversity in the U.S. population has sharpened concerns about the vitality and diversity of the clinical research workforce, concerns that have persisted for two decades. Our nation's unprecedented level of investment in biomedical research has led to an explosion of new knowledge about human health and disease, but basic research achievements must be translated into treatments and therapies in order to benefit human health. This translation requires clinical research conducted by outstanding scientists, physicians, and other health professionals who understand the complexities and nuances of health and disease among different population groups.

Clinical research as an enterprise has traditionally not received the high level of regard afforded basic research in the research and academic communities, which may be contributing to decreased interest in clinical research careers among matriculating medical students. This must change if we are to continue the pace of achievement in translating gains in basic science to treatment of human disease. All biomedical researchers have a stake in ensuring that the clinical research workforce thrives and diversifies for the benefit of human health.

This report has been reviewed in draft form by individuals chosen for their diverse perspectives and technical expertise, in accordance with procedures approved by the National Academies' Report Review Committee. The purpose of this independent review is to provide candid and critical comments that will assist the institution in making its published report as sound as possible and to ensure that the report meets institutional standards for objectivity, evidence, and responsiveness to the study charge. The review comments and draft manuscript remain confidential to protect the integrity of the process.

We wish to thank the following individuals for their review of this report: Karen Antman, National Cancer Institute for Translational and Clinical Sciences; Elaine Gallin, Doris Duke Charitable Foundation; Page Morahan, Hedwig van Ameringen Executive Leadership in Academic Medicine Program; Jay Moskowitz, Pennsylvania State University; Joel Oppenheim, New York University; Diane Wara, University of California, San Francisco; and Judith Woodruff, Northwest Health Foundation.

Although the reviewers listed above have provided many constructive comments and suggestions, they were not asked to endorse the conclusions or recommendations, nor did they see the final draft of the report before its release. The review of this report was overseen by Elena Nightingale, Institute of Medicine, and Willie Pearson, Georgia Institute of Technology. Appointed by the National Academies, they were responsible for making certain that an independent examination of this report was carried out in accordance with institutional procedures and that all review comments were carefully considered. Responsibility for the final content of this report rests entirely with the authoring committee and the institution.

E. Albert Reece, M.D.
Chair

Contents

SUMMARY   1

1   INTRODUCTION   7
Lessons from the Business Sector, 10
Implications for Academic Health Centers, 10
The Focus of This Study, 11

2   THE CLINICAL RESEARCH WORKFORCE:
ACROSS-THE-BOARD CHALLENGES   14
NIH Investment in the Clinical Research Workforce, 15
Workforce Challenges for the Private Sector in Clinical
    Research, 22
The Shortage of Clinical Investigators, 23
Future Needs, 32

3   THE STATUS OF WOMEN AND UNDERREPRESENTED
MINORITIES AND PROGRAMS OF SUPPORT   33
Women Faculty, 33
Women Medical School Students, 37
Underrepresented Minority Faculty, 37
Underrepresented Minority Students in Medical Schools, 39
NIH Programs for Clinical Research and Minority Researchers, 42

Department of Veterans Affairs Programs, 47
Private Sources of Funding for Clinical Investigators, 49
Future Directions, 52

4 THE STATUS AND FUTURE ROLE OF ACADEMIC
   NURSING IN CLINICAL RESEARCH   55
   The Advancing Age of Nursing Faculty, 55
   Preparing a Diverse and Representative Clinical Research
      Workforce, 57
   National Institute of Nursing Research, 59
   Future Needs at the Interface of Nursing and Clinical Research, 61

5 CONCLUSIONS AND RECOMMENDATIONS   66
   Recommendations, 68

REFERENCES   75

APPENDIXES

A Biographies of Speakers   87
B Workshop Guests   95
C Workshop Agenda   102
D Public Mechanisms for Clinical Research Training:
   Examples of Minority Research Training Programs   107
E Public Mechanisms for Clinical Research Training   116
F Examples of Pharmaceutical Company Training Programs   122

List of Tables, Figures, and Boxes

TABLES

2-1 NIH Clinical Research Awards, FY 1996-FY 2001, 16
2-2 First-time NIH Applicants and Awards, FY 1995-FY 2001, 17
2-3 M.D. and Ph.D. NIH Applications, Awards, and Success Rates, FY 1990-FY 2001, 17
2-4 Targeted NIH Clinical Research Awards (Type 1—K23, K24, and K30), FY 1999-FY 2003, 18

3-1 Distribution of Full-Time U.S. Medical School Faculty by Sex and Rank, 2003, 34
3-2 Hispanic Ethnicity and Non-Hispanic Race Medical School Applicants by Acceptance Status, 2002 and 2003, 40
3-3 Distribution of Loan Repayment Program Applicants by Sex, FY 2003, 46

4-1 Race and Ethnicity of Graduates from Baccalaureate, Master's, and Doctoral Programs in Nursing, 1999-2002, 58

FIGURES

1-1 Percent of the population by race or ethnicity: 1990, 2000, 2025, and 2050, 8

3-1 Black, Native American, and Hispanic U.S. medical school faculty, 1980-2000, 38
3-2 Medical school faculty by race/ethnicity, 2002, 38
3-3 Black, Asian, and Hispanic M.D.–Ph.D. graduates, 1986-2002, 42
3-4 New applications and funded awards for four NIH loan repayment programs, FY 2002 and FY 2003, 45

BOXES

2-1 Recommendations of the 2003 NIH Director's Blue Ribbon Panel on the Future of Intramural Clinical Research, 20

3-1 Summary, 53

4-1 Summary, 63

Summary

The increasing diversity and age of the U.S. population present new challenges for the U.S. clinical research community, whose role is to develop healthcare therapies and paradigms from the knowledge gained in basic research. A particularly acute challenge is the need to replenish and diversify its workforce, especially physician-scientists and nurses, whose small numbers are insufficient to meet the increasing need for clinical research. This project aimed to identify ways to recruit and retain more women and underrepresented minorities into the clinical research workforce to meet these challenges.

The study described in this volume incorporated a review of the current state of knowledge about the clinical research workforce and an information-gathering workshop of stakeholders—clinical researchers, medical school deans at academic health centers, and sponsors of clinical research. The study committee developed a set of questions to provide guidance to the workshop presenters and stimulate discussion among the participants:

- What is the benefit of increasing the representation of women and underrepresented minorities in the clinical research workforce? Will increased diversity improve delivery of the results of clinical research to minority communities?
- What are the needs of the private and public sectors? Are the current approaches to training clinical investigators meeting the needs of academia, industry, and public health? Where is demand exceeding supply?

- What training programs and career tracks appear to foster the development and retention of women and minorities in the clinical research workforce?
- What research related to evaluation of existing training efforts needs to be funded? What are the important measures of outcome?

FINDINGS OF THE STUDY

The benefits of increased diversity in the clinical research workforce include increased clinical trial accrual of underrepresented minorities, more robust hypothesis generation for research questions relating to women and minority populations, and the potential for improved understanding and application of the results of clinical research to minority communities.

Unfortunately the study scope, as framed by the questions in the study charge, was much broader than that answerable by the available body of data. The committee found that the first three issues in the study charge could not be fully answered because of the lack of data on the clinical research workforce. This absence of data severely limited the ability of the committee to address questions regarding supply and demand and outcome measures for existing training efforts. Data on the private sector workforce are also not available, similarly limiting the committee's ability to address the study charge about the needs of the private sector.

The data collection needed for accurate characterization of the clinical research workforce is limited by the lack of a common definition of clinical research used across all sectors. The use of standard definitions among federal agencies, careful tracking of the subsets of clinical research, and systematic evaluation of the outcomes of existing training efforts would allow better monitoring of the clinical research workforce.

Physicians have less interest in research careers, and fewer trainees are opting for an M.D.-Ph.D. More women are earning their M.D.s, but fewer are opting for research careers despite continuing interest in academic positions. Underrepresented minorities earning M.D.s have increased numerically, but they are an infinitesimal proportion of the historical increase in M.D.s overall. The shortage of nursing faculty severely restricts the training of future nurses for clinical research and practice. Various training programs and career tracks foster the development and retention of women and minorities in the clinical research workforce, but more are needed for significant improvements in this area. Insufficient data on the clinical workforce limit understanding of its supply and demand, and an insufficient evaluation of existing programs limits assessment of success. Interdisciplinary research among basic and clinical scientists would broaden clinical research interest and should be encouraged.

RECOMMENDATIONS

The study committee clustered its recommendations around the following themes:

1. Adequate collection of the appropriate data;
2. Evaluation of the training landscape and mechanisms;
3. The special needs of nursing;
4. The pipeline and the career path for clinical researchers; and
5. The role of professional societies.

These themes contain systemic challenges that affect the entire clinical research enterprise, as well as specific challenges that should be addressed to improve the strength, character, and diversity of the workforce.

Data Needs

A fundamental difficulty in examining issues surrounding clinical research is the lack of data on the clinical research enterprise as a whole, including data on funding levels, training programs, and who participates in the workforce. It is a challenge to examine ways to sustain and replenish the clinical research workforce when the existing data do not permit an understanding of the state of the clinical research enterprise.

Recommendation

The National Institutes of Health (NIH) of the Department of Health and Human Services should initiate a process that will develop the consistent definitions and methodologies needed to classify and report clinical research spending for all federal agencies, with advice from relevant experts and stakeholders (federal sponsors and academic centers). Such a step would allow a better understanding of the training and funding landscape and would enable accurate data collection and analysis of the clinical research workforce.

Training Landscape and Mechanisms: An Evaluation

Clinical research training programs are supported by public (federal government) and private (industry, foundations) sources and are implemented at academic institutions. Continued support is vital to the health of the clinical research workforce, but awareness of and access to the programs are critical if the workforce is to thrive. The effectiveness of programs should be evaluated on a regular basis to determine their efficacy.

Recommendation

The Department of Health and Human Services should work with federal clinical research sponsors to identify and describe all federally sponsored training programs (both institutional and individual) for clinical research. The information provided should identify support for each level of training and each discipline across the spectrum of clinical research. Organized links to these programs should be available on a website, including programs offered at NIH, the Agency for Healthcare Research and Quality (AHRQ), the Veterans Administration (VA), the Centers for Disease Control and Prevention (CDC), and the Health Resources and Services Administration. This resource should also be open to listing the institutional and individual programs offered by private sponsors for clinical research training.

The committee supports the development of the training website offered by NIH (http://www.training.nih.gov/careers/careercenter) and encourages NIH to modify and expand this resource to include a focus specifically on clinical research training programs.

Academic institutions should document and make publicly accessible the available programs for enhancing the participation of women and minority trainees in clinical research.

The sponsors of federal, foundation, and industry clinical research training programs should continue to support the existing efforts to train, develop, and sustain the careers of clinical researchers.

Recommendation

Federal sponsors (NIH, CDC, AHRQ, VA, Department of Defense) should ensure adequate representation of women and minorities in study section review panels that review clinical research.

Recommendation

Federal agencies and academic institutions should periodically evaluate how well their current training programs are enhancing the racial and ethnic diversity of trainees and they should modify these programs as needed to increase the programs' effectiveness in clinical research.

Nursing Professionals

The continuing shortfall of nursing professionals is compounded in clinical research by the longer time required for specialized training, and the fewer numbers of nursing faculty involved in clinical research.

Recommendation

The need for appropriately trained nursing professionals in the clinical research workforce is especially urgent. A significant push is needed to increase the numbers of minorities entering the nursing profession. Additional attention should be paid to the clinical research training of nurse-scientists, nursing students, and nursing faculty at all academic levels.

The shortfall could be curtailed by expanding training efforts. These could include increasing fast-track B.S.N.-Ph.D. programs, training grants in clinical research, summer programs, fellowships, and training sabbaticals.

Replenishing the Pipeline: A Flexible Career Path

Given the long training period required for clinical research, entry points throughout a clinical research career path, not just at trainee levels, could increase the workforce. Additional efforts are needed to retain scientists in the clinical research workforce.

Recommendation

Academic institutions should develop strategies to attract mentors and reward mentorship in clinical research training. A special emphasis should be placed on the women and minorities who carry the greatest burden of mentorship responsibilities for women and minority scientists.

Recommendation

Federal sponsors of clinical research should amplify the existing funding mechanisms and create new ones that allow flexibility in career training, such as second-career programs, reentry mechanisms, and service payback agreements. These programs should be described on the NIH training website. In addition, other entry routes into the clinical research path, including short-term training programs, should be developed.

The Role of Professional Societies

Professional societies play a major role in the scientific community, as publishers of journals, sponsors of awards, and representatives of their scientific community.

Recommendation

Specialty medical and nursing societies should form a new consortium that would assume an enhanced role in fostering a diverse clinical research workforce.

1

Introduction

According to projections of the U.S. Census Bureau, the demographics of the United States population will change dramatically over the next five decades. By 2050 whites will comprise 53 percent of the general population, Hispanics 25 percent, Asians nearly 9 percent, and blacks 15 percent (see Figure 1-1). Females will outnumber males by over 6 million, and the average age of the population will become older, with one in five persons over the age of 65. For the biomedical community these demographic changes present considerable challenges for both research and healthcare delivery.

The increased diversity in the population has not been reflected in the composition of the healthcare and biomedical research workforces, which is an issue of considerable concern to the biomedical and healthcare community. Indeed, the need for diversity in the healthcare workforce was recently examined by the Institute of Medicine (2004a).

If the need for diversity in healthcare delivery is acute, the need in the clinical research workforce is even more so. Before healthcare practices can be developed and introduced into primary care, much research must be conducted, both basic and clinical. Because of the historically lower rates of participation in research by women and ethnically diverse groups, both among the workforce and as participants in clinical trials, the challenge of meeting the complex healthcare needs of an ever more diverse population is particularly difficult.

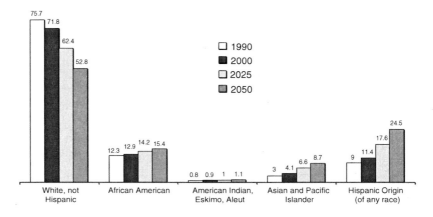

FIGURE 1-1 Percent of the population by race or ethnicity: 1990, 2000, 2025, and 2050.
SOURCE: U.S. Census Bureau, decennial census and population projections.

The need for greater diversity in the clinical research workforce is compelling. Racial and ethnic minority healthcare professionals are more likely to serve minority populations (Cantor et al., 1996; Komaromy et al., 1996). Minority patients are more likely to select healthcare professionals with their own ethnic background (Saha et al., 1999). Consequently, healthcare professionals from racial and ethnic minority groups may be more successful in recruiting minority patients to participate in clinical research. Such efforts are critical to linking scientific advancements with quality service and delivery in underserved minority communities.

In some ways minority researchers may be better positioned to formulate the right research questions as well as to devise ways to answer them. When compared with the majority population, minority populations in the United States experience higher rates of disease and mortality (e.g., cancer, cardiovascular disease, diabetes, HIV/AIDS, infant mortality), and they are less likely to receive regular, high-quality medical and preventive healthcare services (NIH, 1994; Corbie-Smith et al., 1999; Giuliano et al., 2000; Killien et al., 2000; Gifford et al., 2002). Specifically, black men are more likely to be diagnosed with prostate cancer. Asian Americans are more likely to get stomach and liver cancer. The American Indian population has the lowest cancer survival rates of all (Haynes, 1999). Some of these disparities can be attributed to socioeconomic differences and poorer access to health insurance. However, access to quality health care for these populations may also be affected by the diversity of the healthcare and clinical research workforce (NIH, 1994; Corbie-Smith et al., 1999; Giuliano et al., 2000; Killien et al., 2000; Gifford et al., 2002).

Gender adds yet another dimension to an already complex problem. Since the establishment of the Office of Research on Women's Health at the National Institutes of Health, a tremendous amount of information has been gained. In biomedical research, gender is clearly a critical factor in understanding human health (IOM, 2001a). Minority women and white women experience different rates of disease. Among Hispanic and Vietnamese American women, cervical cancer rates are higher. African American women are also less likely to survive breast cancer, although they are less likely than others to develop it (Haynes, 1999).

Women are critical to clinical research not only as participants in clinical trials but also as researchers. A driving factor in the need to recruit women into the clinical research workforce is that they are likely to be the majority of M.D. recipients in the future and therefore the pool from which researchers must be drawn. In the past few decades the number of M.D.s awarded to women has steadily increased; in 2003 females accounted for almost 50 percent of medical school enrollment (AAMC, 2003). In the basic sciences women currently receive half of the bachelor's degrees issued in the biological sciences and over 40 percent of the Ph.D.s (NSF, 2004), and the trend is toward continued increases in their proportions of these degrees.

Clearly then, women and underrepresented minorities are crucial to replenishing the clinical research workforce. A diverse workforce in the sciences leads to many benefits—among others, a wide diversity of perspectives leading to better opportunities for scientific advancement, and a potentially intensified focus on understanding and eradicating health disparities among different ethnic and racial groups (Crowley et al., 2004). Research indicates that cultural differences are often at the core of miscommunication and dissatisfaction in the physician–patient relationship; culture also can significantly influence patient health outcomes (Anderson, 1995; Airhihenbuwa et al., 1996; Berger, 1998; Hunt et al., 1998). Moreover, diverse teams can outperform homogeneous ones (Lippman, 2000; Sessa and Taylor, 2000). Managers who are exposed to professionally and culturally diverse colleagues cultivate new ideas by drawing on a larger pool of information and experiences.

LESSONS FROM THE BUSINESS SECTOR

Researchers in the business sector have learned that a diversified staff facilitates marketing to a more diversified customer base, which increases market share (Allen and Montgomery, 2001). They have also learned that companies with reputations for good diversity management are more successful in attracting and retaining top-quality employees (Ferraro and Martin, 2000). Likewise, companies with high ratings on equal employment opportunities outperform those with poor ratings on hiring and advancing women and minorities (Adler, 2001). *Fortune* 500 companies with the highest percentages of women executives deliver earnings far in excess of the median compared with large firms with the fewest women. Among initial public offerings, companies with women in senior management received higher valuations and had better long-term performance (Church, 2001).

Private sector companies have thus begun to recognize that diversity is associated with enhanced productivity and lower turnover costs among highly trained employees. The economic advantages of a diverse workforce are even greater for businesses that serve a diverse clientele (McCracken, 2000). In 1991 the accounting firm Deloitte and Touche was experiencing a high rate of attrition among women professionals, which company leaders initially attributed to societal reasons. The realization among those leaders that a sizable share of the company's primary product—its talent—was leaving each year led to cultural changes at Deloitte and Touche that have been widely regarded as successful. During the nine years after the implementation of an initiative for retention and advancement of women, the proportion of women full partners and directors increased from 5 percent to 14 percent, and attrition rates for men and women equalized. In addition, overall retention rates improved substantially, which saved an estimated $250 million in hiring and training costs and has supported increased productivity among the retained staff (Mueller, 1998).

IMPLICATIONS FOR ACADEMIC HEALTH CENTERS

What are the implications of these "lessons" for academic health centers? A recent study concluded that academic health centers (AHCs) that benefit from women's intellectual capital receive both short- and long-term payoffs (Morahan and Bickel, 2002). Female patients are seeking female surgeons and subspecialists. Likewise, students are seeking female role models in these fields (Bickel, 2001; Morahan and Bickel, 2002). As the proportion of women students continues to increase, only those institutions able to recruit and retain women in all departments will have the best house staff, faculty, and administrators (Cox, 1993). The absence of women in key positions can be a negative signal to female candidates.

A growing number of studies indicate that positive academic and social outcomes for students can be attributed to diversity in higher education that benefits the training of health professionals (IOM, 2001b). Bowen and Bok (1998) studied the educational and career outcomes of two cohorts of majority and minority students attending selective colleges and universities. They found that the minority graduates of these institutions attained levels of academic achievement that were on a par with those of their nonminority peers. More specifically, these minority graduates obtained professional degrees in fields such as law, medicine, and business at rates far higher than the national averages for all students (Bowen and Bok, 1998). Similar findings were obtained in a study of the academic outcomes of college students attending racially and ethnically diverse colleges and of those attending less diverse institutions. Gurin (1999) concluded that students can best develop the capacity to understand the ideas and feelings of others in an environment characterized by a diverse student body, equality among peers, and discussion of the rules of civic discourse.

Although it should not be assumed that women and other underrepresented individuals would choose research topics directly related to their populations, their enhanced representation and involvement may improve perspective and understanding in hypothesis generation for all research in which they are involved. Furthermore, those who belong to a group of interest are more likely to have personal experience that will aid in the selection of testable hypotheses and methods appropriate to the population. The participation of minority groups in clinical trials is enhanced by the participation of minority clinical research investigators (NIH, 2002b).

THE FOCUS OF THIS STUDY

To examine this issue the National Research Council's Committee on Opportunities to Address Clinical Research Workforce Diversity Needs for 2010, supported in part by the Office of Research on Women's Health at the National Institutes of Health, conducted a study of opportunities to enhance participation and promote diversity in the clinical research workforce. The committee developed the study charge, which was to highlight new paradigms in clinical research and research training (interdisciplinary research and team science) and to issue recommendations for improvements in the training of nurse and physician clinical research investigators. In its work the committee focused on the following questions:

- What is the benefit of increasing the representation of women and underrepresented ethnic groups in the clinical research workforce? Will increased diversity improve delivery of the results of clinical research to minority communities?
- What are the needs of the private and public sectors? Are the current approaches to training clinical investigators meeting the needs of academia, industry, and public health? Where is demand exceeding supply?
- What training programs and career tracks appear to foster the development and retention of women and minorities in the clinical research workforce?
- What research related to the evaluation of existing training efforts needs to be funded? What are the important measures of outcome?

To address these questions the committee gathered information from numerous sources and held a workshop of stakeholders—clinical researchers, academic health center deans, and funders of clinical research in 2003. The information gathering was directed toward assessing current progress on increasing the participation of women and underrepresented minorities in clinical research and identifying workforce sectors that require attention (see Appendix A for biographies of the workshop speakers, Appendix B for a list of workshop participants, and Appendix C for the workshop agenda).

The workforce and training needs for all biomedical research, including clinical research, have been monitored by the National Institutes of Health since 1975. As mandated by Congress, the National Research Council has conducted an ongoing assessment of the nation's overall need for biomedical and behavioral research personnel, the subject areas in which researchers are needed and the numbers of personnel required in those areas, and the type of training needed by researchers. The original study committee interpreted the charge to include clinical research, and it monitored clinical research scientists and training. The monitoring of clinical research has continued throughout the series of reports issued in conjunction with the study.

Each study report in the series has cited the challenges encountered in trying to estimate the workforce and training needs for the clinical research

INTRODUCTION

community. The primary obstacle cited is the dearth of data on the clinical research workforce.

The committee for the study described in this volume agreed with the biomedical research workforce study reports that the lack of a clear, agreed-upon definition for clinical research is a significant obstacle to the collection of data. Based on this and other information, the current study committee concentrated its effort on the recruitment of M.D.s, M.D./Ph.D.s and Ph.D. nurses into clinical research.

The study committee found that clinical research presented a set of challenges different from those posed by basic science for students considering research careers. Chapter 2 provides a framework for understanding the systemic challenges facing those who educate, fund, and employ clinical researchers. Chapter 3 outlines the current status of women and underrepresented minorities in clinical research, as well as the programs that have been devised to aid their advancement. The specific needs of nursing professionals are addressed in Chapter 4. Chapter 5 presents the study's conclusions and recommendations.

2

The Clinical Research Workforce: Across-the-Board Challenges

Over the past two decades policy makers, researchers, and others have consistently expressed concern about the state of the clinical research workforce (Wyngaarden, 1979; Ahrens, 1992; IOM, 1994; NIH, 1997; Zemlo et al., 2000). At the same time, expectations have grown about possibilities of translating new life sciences knowledge into health benefits (NIH, 1997; Zemlo et al., 2000; Elliott, 2001; Fey, 2002; Bloom, 2003; Gerling et al., 2003; Mike, 2003).

Because the challenges of a clinical research career are very different from those of a basic science career, the study committee believed a general overview of clinical research would be helpful in understanding how these challenges particularly affect women and underrepresented minorities. Hence, this chapter will provide a view of the challenges facing the clinical research workforce overall. This includes an overview of the efforts by the National Institutes of Health (NIH) to promote clinical research over the past decade, as well as a discussion of the clinical research workforce challenges of the private sector, a major sponsor of clinical research.

As noted in Chapter 1, since 1975 NIH has monitored the workforce and training needs for all biomedical research, including clinical research (NRC, 2005a). The reports issued in conjunction with this monitoring have pointed out that the primary obstacle to estimating workforce and training needs for the clinical research community was the lack of data on the clinical research workforce.

NIH INVESTMENT IN THE
CLINICAL RESEARCH WORKFORCE

In 1994 the Institute of Medicine (IOM) issued a report on careers in clinical research that focused on three major issues: (1) accurate data on the numbers of clinical researchers, (2) career training support and funding, and (3) a systematic review of research administration and infrastructure (IOM, 1994). In 1995 NIH convened a director's advisory panel to examine the challenges facing clinical research, such as financing, the role of clinical research centers, the recruitment and training of its workforce, and the conduct of clinical trials and peer review.

The actions produced by the recommendations of the advisory panel were instrumental in advancing clinical research in three ways. First, NIH adopted definitions for clinical research that allowed better collection of data, and it constructed a landscape view of who had taken up careers in clinical research (NRC, 2000). Second, NIH examined the composition and outcomes of study sections to ensure that clinical research proposals were being reviewed by those with clinical expertise. Third, NIH developed mechanisms for the training and support of clinical investigators (e.g., a series of K awards and the Clinical Research Loan Repayment Program).

More recently NIH embarked on an ambitious new plan for medical research in the twentieth-first century, the Roadmap for Accelerating Medical Discovery to Improve Health, which features a major emphasis on clinical research. Re-engineering the Clinical Research Enterprise aims to facilitate the bench-to-bedside transition through, among other things, enhanced regional clinical research centers that incorporate academic health centers as well as community-based healthcare providers, better organization of the gathering of clinical research information, better information technology, and ways to enhance the workforce. Within NIH itself, a panel to examine intramural clinical research has been established to provide guidance and review. (Intramural research is conducted within NIH laboratories. Extramural research is conducted by researchers at academic institutions that receive grants from the NIH.)

NIH Director's Panel on Clinical Research (1996): Status

NIH has not changed the proportion of clinical research support since the launch of the NIH Director's Panel on Clinical Research in 1996.[1] By

---

[1] William Crowley Jr., M.D., workshop presentation, 2003.

TABLE 2-1 NIH Clinical Research Awards, FY 1996-FY 2001

Fiscal Year	Total Competing Awards		Clinical Research Awards		Percent of Total Clinical Research	
	Number	Amount ($ millions)	Number	Amount ($ millions)	Awards	Funding
1996	10,493	2,361	2,795	906	27	38
1997	11,592	2,572	2,767	877	24	34
1998	11,780	2,984	2,882	1,000	24	34
1999	13,971	3,946	3,470	1,257	25	32
2000	15,357	5,278	3,862	1,722	25	33
2001	14,622	4,717	3,874	1,609	27	34

SOURCE: NIH (2002a).

2001, NIH grants had increased by almost 50 percent, representing more than a 50 percent increase in their dollar value (see Table 2-1). The increase in clinical research grants has been roughly comparable to the increase in total competing awards.

M.D. or Ph.D. rates of applications have seen slight improvements. The number of awards has increased substantially for first-time M.D. applicants, but as of 2001 there had been no significant change in the number of applicants (see Table 2-2). The same is true for Ph.D.s. Between 1997 and 2001 the overall M.D. success rate increased to between 32 percent and 35 percent (see Table 2-3).

The average growth rate for M.D. applicants was only 2.30 percent versus 4.05 percent for Ph.D.s. A concern is whether this flat rate of growth will ensure an adequate supply of M.D. applicants in an increasingly clinic-oriented research environment.

The renewal rate for RO1-funded (RO1's are research grants awarded to independent investigators at academic institutions) clinical investigators has been lower than that of nonclinical research grantees. A little over 30 percent of all nonclinical investigators who received awards in 1996 and reapplied in 1999-2001 were renewed, whereas only 17 percent of clinical research investigators were renewed. Of the 405 clinical researchers who applied for awards in 1966, only 47 percent sought renewal, versus the 68 percent of the 954 nonclinical researchers who sought awards in 1996.[2] Although targeted clinical research career awards have been successful, the

TABLE 2-2 First-time NIH Applicants and Awards, FY 1995-FY 2001

	M.D.s			Ph.D.s		
Fiscal Year	No. of Applicants	No. of Awards	Success Rates (%)	No. of Applicants	No. of Awards	Success Rates (%)
1995	2,150	385	17.9%	5,136	982	19.1%
1996	1,930	399	20.7%	4,802	952	19.8%
1997	1,915	408	21.3%	4,671	1,042	22.3%
1998	1,876	425	22.7%	4,452	1,059	23.8%
1999	1,965	489	24.9%	4,853	1,052	21.7%
2000	1,941	491	25.3%	4,892	1,078	22.0%
2001	1,962	470	24.0%	4,893	1,055	21.6%

SOURCE: Nathan and Wilson (2003).

TABLE 2-3 M.D. and Ph.D. NIH Applications, Awards, and Success Rates, FY 1990-FY 2001

	M.D.s			Ph.D.s		
Fiscal Year	No. of Applicants	No. of Awards	Success Rates (%)	No. of Applicants	No. of Awards	Success Rates (%)
1990	5,707	1,484	26.0%	15,281	3,683	24.1%
1991	5,784	1,677	29.0%	15,112	4,337	28.7%
1992	5,705	1,740	30.5%	15,575	4,548	29.2%
1993	6,408	1,512	23.6%	16,638	3,960	23.8%
1994	7,283	1,792	24.6%	17,646	4,570	25.9%
1995	7,004	1,891	27.0%	17,680	4,738	26.8%
1996	6,338	1,787	28.2%	16,979	4,771	28.1%
1997	6,515	2,058	31.6%	17,214	5,164	30.0%
1998	6,513	2,143	32.9%	17,162	5,252	30.6%
1999	7,481	2,672	35.7%	18,382	5,775	31.4%
2000	7,925	2,712	34.2%	19,041	5,896	31.0%
2001	7,998	2,839	35.5%	19,816	6,137	31.0%

SOURCE: Nathan and Wilson (2003).

TABLE 2-4 Targeted NIH Clinical Research Awards (Type 1: K23, K24, and K30), FY 1999-FY 2003

Fiscal Year	K23	K24	K30	Total
1999	85	81	35	201
2000	193	77	22	292
2001	185	58	0	243
2002	194	48	2	244
2003	214	40	0	254
Total	871	304	59	1,234

SOURCE: Walter Schaffer, NIH IMPAC-II, October 7, 2003.

number of K24s has decreased consistently. The loan relief programs and K30s have been critical in relieving medical student debt. The K30 programs meet educational needs, but they are for people still in training, not for independent investigators (see Table 2-4).

NIH Director's Blue Ribbon Panel on the Future of Intramural Clinical Research

The Blue Ribbon Panel on the Future of Intramural Clinical Research[3] was convened in August 2003 by NIH Director Zerhouni in response to three events: (1) the building of the new Clinical Research Center (CRC), (2) the NIH roadmap, and (3) changing approaches in academic health centers (AHCs) to clinical research. The following questions were posed to the panel:

1. In what areas not addressed by other NIH-supported research can the Intramural Clinical Research Program (ICRP) produce paradigm-shifting research?
2. Is the current ICRP portfolio suitable?
3. How can the ICRP enhance the overall NIH-supported clinical research enterprise?

---

[2]Ibid.
[3]See http://www.genome.gov/Pages/About/NACHGR/2004NACHGRAgenda/Tab%20H%20-%20BlueRibbonPanel.pdf. Date accessed October 19, 2004.

4. How can this be accomplished by reassigning existing resources to excellent, distinctive intramural programs in a steady-state environment?
5. What measures should be used to assess the success of the ICRP?

Based on the charges to the panel, recommendations were made to promote the status of clinical research within the NIH enterprise (see Box 2-1); NIH developed responses to the recommendations.

- The ICRP should adopt streamlined, comprehensive governance. Rather than being an impediment to innovation, the governance structure should help the clinical research enterprise to realize its full potential.
- The career pathways of patient-oriented research should be strengthened and rewarded. Creating a clear and rewarding career path for clinical investigators is an essential first step toward attracting and retaining clinical investigators committed to conducting patient-oriented research.
- NIH could champion the development of new therapies for several rare diseases in order to potentially affect the economics of drug development and serve an unmet medical need. By focusing on diseases with which NIH investigators have special expertise, the intramural program could enhance downstream clinical investigation.
- Clinical research should have general clinical research center (GCRC) and children's clinical research center (CCRC) multicenter networks. Improvements in the infrastructure will be needed to revitalize the roles of GCRCs and CCRCs in clinical research and establish them as centers of innovation.
- A focus on research that defines a more distinctive niche in the U.S. biomedical research portfolio and takes on projects that cannot be included in the extramural program should be included in future directions. The intramural program has fewer short-term constraints than the extramural program and is particularly well equipped for the conduct of innovative and bold research.
- Streamlining, regulatory reform, and standardization are needed. Any steps to encourage investigators to initiate new protocols and harmonize the demands placed on clinical research from different regulatory agencies will require making the regulatory and review processes for intramural research more efficient and uniform across institutions.

**BOX 2-1
Recommendations of the 2003 NIH Director's
Blue Ribbon Panel on the Future of
Intramural Clinical Research**

1. Revise the NIH intramural clinical research oversight structure.

   - Create a single, high-level oversight committee to replace all existing governing bodies that have oversight responsibilities for intramural clinical research.
   - Create an external advisory committee that reports to the NIH director to periodically and systematically consider the overall quality and vitality of the NIH Intramural Clinical Research Program.
   - Strengthen the roles of the Office of the Director and institute and center leadership in clinical research.

2. Develop new training and career pathways in patient-oriented research.

   - Strengthen career pathways and mentoring in the ICRP for patient-oriented research that culminate in tenure.
   - Establish a premier, highly visible postdoctoral fellowship program in clinical research, administered by the CRC director, for individuals who have finished clinical residency training.
   - Create an advanced research training program for extramural faculty members in AHCs who wish to take sabbatical at the CRC as a means of obtaining on-the-job experience in clinical research.
   - Foster the recruitment and retention of innovative patient-oriented investigators in the ICRP by ensuring that salaries and benefits are competitive with those at AHCs.
   - Foster an interactive and creative research environment that will attract outstanding postdoctoral fellows. Postdoctoral fellows will want to participate in programs that are conducting disease-oriented research or investigating timely clinical problems that cannot easily be studied in extramural AHCs.

3. Continue to emphasize the study of rare diseases at the CRC and promote a strong emphasis on pathophysiology and novel therapeutics in the ICRP.

    - Initiate trans-NIH programs of patient-oriented research that combine the expertise of several institutes and centers.
    - Make the best use of the unique features of NIH's intramural research program and its ability to undertake bold and innovative research.

4. Create clinical, multidisciplinary intramural and extramural partnerships involving the GCRCs, CCRCs, NIH-funded extramural networks, the CRC, and the ICRP.

5. Ensure that Intramural clinical research, including new programs in patient-oriented investigations, is excellent and distinctive, as well as distinguishable from research conducted at AHCs.

    - This mandate for change should be the responsibility of the NIH director, institute and center leaders, the advisory committees, and the basic science centers.

6. Reduce regulatory barriers and impediments to clinical research. This would include streamlining the regulatory process and providing adequate, effective infrastructure for supporting clinical research.

SOURCE: NIH (2004).

WORKFORCE CHALLENGES FOR THE PRIVATE SECTOR IN CLINICAL RESEARCH

As innovators in drug development, the pharmaceutical industry has always been an active participant in clinical research. In 2002 the total worldwide expenditures by major pharmaceutical companies for research and development (R&D) were $44.5 billion, of which 62 percent was devoted to pre-approval (Phases I-III) and 38 percent to Phases IIIb-IV (PhRMA, 2004).[4] Industry maintains a steady proportional investment in R&D—18 percent of sales. In the United States, Phase I spending in 2002 exceeded $1.8 billion; spending on Phase II and Phase III clinical research projects was $2.3 billion and $3.6 billion, respectively; and Phase IV spending reached $1.5 billion (Thomson CenterWatch, 2004).

It is not uncommon for Phases I-III of a drug development program to involve as many as 10,000 or more patients before a drug is approved. If investigators want to host drug development programs and patients are in relatively short supply, many different centers do the research, that is, from 100 to 1,000 research centers around the world recruit the patients needed for a particular clinical development program. Patient recruitment and retention remain the largest problems in drug development, despite significant increases in spending to reach and randomize study volunteers (Thomson CenterWatch, 2004). As a result, clinical research has moved away from small focused studies within academic institutions to large multicenter trials, including more community physicians (Phillips, 2000; Randal, 2001; Gelijns and Thier, 2002).

Recruiting Clinical Investigators

Some 100,000 physicians are in training in the United States, which is the beginning of the clinical research pipeline. From this pool 1,700 physicians were hired in 2002 to full-time AHC clinical positions. The 87,000 clinical faculty in academic health centers are an important source for

---

[4]"R&D expenditures" are defined as expenditures within PhRMA member companies' U.S. and or foreign research laboratories plus R&D funds contracted or granted to commercial laboratories, private practitioners, consultants, educational and nonprofit research institutions, manufacturing and other companies, or other research-performing organizations. "Phases I/II/III clinical testing" is defined as from first testing in designated phase to first testing in subsequent phase. "Phase IV clinical testing" is defined as any postmarketing testing performed.

industry recruitment. The 2.1 percent increase in this faculty from 1999 to 2000 indicates that the talent pool is growing (Barzansky and Etzel, 2002).

Because only 2 percent of research worldwide occurs within pharmaceutical companies (Yates, 2003), the private sector frequently collaborates with AHCs (and smaller companies) in conducting its clinical research. However, the majority of resident physicians-in-training do not receive formal exposure to research methods as part of their clinical development. Even fewer receive exposure to regulatory good clinical practice training, which is the core competency for clinicians working in industry (Martinez, 2003). Facing increasingly strained finances, academic institutions do not allocate sufficient resources to clinical research and the faculty necessary to carry it out well. In turn, the physician time devoted to clinical investigation or to experimenting with the development of novel healthcare delivery approaches has diminished (Snyderman, 2004).

Lack of formal standardized training, appropriate certification, and adequate time for research are some of the limits placed on clinical researchers (Snyderman, 2004). Although the incentives offered by certifications and advanced research master's and Ph.D. degrees should not be overstated. These incentives are not nearly as important as experience and exposure to these methodologies, which could be incorporated into the medical school curriculum. General clinical training could be made more accessible in a number of ways; for example, Web-based clinical research training programs could be offered so that continuous on-demand access would be available. Such programs could provide continuing medical education (CME) credits and course certification on good clinical practices (GCPs) and the processes of doing research. Starting and maintaining a research career demand a great deal from young physicians; the acquisition of additional degrees is not promoted (Martinez, 2003).

To encourage physicians to pursue careers in clinical research, several pharmaceutical companies sponsor training in that research (examples are listed in Appendix H). Such training awards range from summer programs to postdoctoral training. Some fellowship programs are specifically targeted to disease areas. Information about these programs is not centrally located, and individual sponsors must be contacted for award eligibility details.

THE SHORTAGE OF CLINICAL INVESTIGATORS

In a trend that parallels the growing need for clinical study participants, a shortage of adequately trained clinical investigators may begin to develop (Goldman and Marshall, 2002). One reason for this shortage lies in the complex and incompletely defined factors that are influencing the career choices of medical students. The many basic science; clinical care; and social, ethical, and economic issues addressed by today's medical school curricula leave less time to educate students about the significance of biomedical research in bettering health care or to inspire students to participate in biomedical research. Exposure to research early in medical school training encourages involvement in and provides a basis for the student to pursue research training (Solomon et al., 2003). Five major obstacles dominate the clinical research landscape:

1. Educational debt;
2. Length of clinical research training time;
3. Perceived challenges to recognition and promotion of clinical researchers;
4. Inadequate training in clinical research techniques; and
5. Increased regulations and monitoring of clinical research.

At each stage of a clinical research career the associated obstacle can serve as a significant deterrent, or at least steer a potential researcher into a different career path; for example, the median debt among graduates of public and private medical schools is $70,000 and $100,000, respectively (Heinig et al., 1999). Before achieving independence, investigators may spend 5-10 postgraduate years in training (Wolf, 2002; Sung et al., 2003). Tenure and promotion standards may put investigators who conduct prolonged but noteworthy studies at an academic disadvantage. Significantly increased regulation and monitoring of clinical research can be found at many levels, including at the institutional (AHCs) level and at the federal level (such as at NIH), the Food and Drug Administration (FDA), and the Office for Human Research Protections (OHRP).

Currently only 8 percent of principal investigators conducting industry-sponsored clinical trials are younger than 40 years, and data show inadequate numbers of new investigators to replace the older generation (Goldman and Marshall, 2002). Likewise, less than 4 percent of competing research grants awarded by NIH in 2001 were awarded to investigators aged 35 years or younger (Chan et al., 2002).

In a survey assessing the health and quality of the clinical research enterprise as perceived by AHCs, 75 percent of respondents reported a moderate to large problem in recruiting clinical researchers who were properly trained (Campbell et al., 2001). Two-thirds of respondents among the most research-intensive institutions reported difficulties in recruiting clinical researchers.

The rest of this section discusses the reasons behind the specific shortages of physician-scientists, nurse-investigators, Ph.D.s in clinical research, and other investigators.

Physician-Scientists

The numbers of physician-scientists in the clinical research community are dwindling. A study by Newton and Grayson (2003) reviewing trends in career choice by graduates of U.S. medical schools found that there has been a decreased interest in research careers in both sexes. During the past decade, the percentage of U.S. M.D.s interested in exclusive or significant research careers has decreased by approximately 16 percent (AAMC, 2003). In 2002 only 0.9 percent of medical school graduates received combined M.D.-Ph.D. degrees, down from 2.3 percent five years earlier (NRMP, 2003). For future research in fields that integrate clinical and basic sciences, this trend has obvious implications (Newton and Grayson, 2003).

Indeed, the academic medical community has become increasingly concerned about the challenges facing the clinical research enterprise in the United States. The survey by Campbell et al. (2001) found clinical research in academic health centers to be of poorer quality, less robust, and facing greater challenges than nonclinical basic research. The policies and mechanisms needed to address challenges facing the clinical research mission were not present at many AHCs. Of the AHCs that had such policies, more than half believed that those policies had not had large positive effects. The findings of the survey indicated that the infrastructure and workforce of clinical research might have to be strengthened and expanded to keep up with basic research advances. The sections that follow describe some of the deterrents to such an expansion: the debt burden faced by young M.D.s, physicians' lack of success at obtaining research funding, physicians' lack of mentors, and the disadvantages faced by physician-researchers in gaining promotions.

Debt Burden

Financial pressure may be a driving force in deterring physicians from clinical research careers (Wolf, 2002). Eighty-five percent of graduates of medical school incur significant educational debt (Heinig et al., 1999; AAMC, 2003). According to the Association of American Medical Colleges (AAMC), since 1984 the median tuition and fees have increased by 165 percent in private medical schools and by 312 percent in public medical schools (AAMC, 2004a). In constant dollar terms, the increases have been 50 percent and 133 percent, respectively. From academic year 2002-2003 to 2003-2004, the median tuition and fees increased by 5.7 percent in private schools (3.4 percent in constant dollars) and 17.7 percent in public schools (15.1 percent in constant dollars). In six public medical schools the increases in tuition and fees exceeded 45 percent.

It is not surprising then that today most clinical research physicians and dentists begin their professional careers with sizable educational debt (NRC, 2000). The average medical school debt of M.D. graduates increased more than 50 percent from 1990 to 1997, from almost $41,000 to just over $64,000, and reached an average of $102,000 in 2003.[5] Research training and early career development add extra years, and additional financial pressure is put on all trainees, even those with minimal or no debt.

Obtaining Research Funding

Another deterrent to the entry and retention of physician-scientists in clinical research is their lack of sustained success in securing research funding (DePaolo and Leppert, 2002). In study sections in which both basic and clinical research grants are reviewed, clinical research applications fare less well in peer review than their basic science counterparts (Kotchen et al., 2004). The success rate for NIH research grants from 1996 through 2001 for first-time Ph.D. applicants was higher than that for first-time physician applicants, and the success rate for established investigators was higher than that for first-time applicants (Nathan and Wilson, 2003). M.D. applicant trends at NIH suggest that the research careers of many physicians end with the rejection of their first federal grant application (Wolf, 2002).

Scarcity of Mentors

One vital ingredient in the success of all physicians, including clinicians, basic scientists, and clinical researchers, is superior mentoring. Ideally,

---

[5]See http://www.aamc.org/data/gq/allschoolsreports/2003.pdf. Date accessed December 5, 2004.

mentors direct trainees toward promising educational opportunities; they serve as advocates of trainees; and they lend expertise and funding to trainees' mentor-guided studies. When asked to identify the most useful and positive aspects of their training, recent graduates of medical schools and training programs gave "outstanding mentorship" as their second most common response. The scarcity of experienced mentors and role models was often cited as a disincentive for entering a career in clinical research (AAMC, 1999). Today fewer capable mentors are available because fewer clinical researchers were trained in the past. Academic institutions' mission to train the next generation of clinical scientists will erode if this trend is not reversed (Wolf, 2002). A study by Buckley et al. (2000a) found less time, less mentoring, and fewer resources for an academic career available to physician faculty who spent the majority of their time in clinical activities.

Receiving Academic Promotions

The promotion standards of academic institutions are uniform across all types of research, despite the slower pace of clinical research (Wolf, 2002). Investigators who choose to undertake essential but lengthy studies are at a disadvantage in receiving academic promotions as a consequence. Survey data reveal that the promotion standards for medical school faculty are centered primarily on research productivity (Beasley et al., 1997). For M.D. faculty with the rank of instructor and above in one institution, the adjusted odds of being satisfied with their progress at promotion were 61 percent lower among clinical researchers than among basic researchers (Thomas et al., 2004). For academic clinicians, the odds of satisfaction were 92 percent lower and for teacher-clinicians 87 percent lower. When academic clinicians and teacher-clinicians were compared with basic research faculty, the adjusted odds of being at a higher rank were found to be 85 percent lower for academic clinicians and 69 percent lower for teacher-clinicians.

The lower growth of M.D.s funded in the clinical sciences by NIH compared with that of M.D.-Ph.D.s and Ph.D.s, together with the declining proportion of NIH award holders less than 45 years of age, indicate that the number of young physician-scientists will decline (Zemlo et al., 2000).

Nurse-Investigators

It is estimated that by 2020 the United States will be experiencing a 29 percent deficit in nursing personnel (IOM, 2004b). This shortfall will have a particular impact on clinical research teams, which often rely on bedside nurses to collect data.

The nursing profession has seen little growth in the number of underrepresented racial and ethnic minorities entering its ranks in recent years (IOM, 2004a). A recent study by Mateo and Smith (2003) of hospital-based nurses and diversity initiatives, outcomes, and issues related to patients and staff found that most respondents did not make management of diversity a priority for the workforce they were managing. In graduate nursing programs underrepresented minority students constituted 12.4 percent of students in master's programs and 8.1 percent in doctorate programs (AACN and the National Organization of Nurse Practitioner Faculties, 2000). Among nursing graduates who were awarded degrees in 2002, White nurses constituted the largest percentage of graduates in baccalaureate, associate, diploma, and R.N. programs, earning between 60 percent and 70 percent of diploma, associate, basic B.S.N., and all basic R.N. degrees (National League for Nursing, 2003). In baccalaureate nursing programs underrepresented minority student enrollment increased by 48 percent between 1991 and 1999, from 11,661 to 17,303 baccalaureate nursing students (National League for Nursing, 2003). Of the U.S. nursing schools listed on the Nursing Spectrum Web site,[6] the majority have a minority affairs office, a diversity center, or some other such entity to recruit and retain minority faculty and students, perhaps contributing to the increase in underrepresented minority student enrollment in baccalaureate nursing programs. Despite this development, underrepresented students, compared with the typical Caucasian or Asian students, are less likely to be enrolled in biological or life sciences or in health profession (nursing and other nonphysician) undergraduate programs (IOM, 2004b).

The shortage of nurses stems from two factors, among others, described in the remainder of this section: (1) a diminishing nursing faculty and (2) an aging R.N. workforce.

---

[6]See http://nsweb.nursingspectrum.com/Education/. Date accessed November 22, 2004.

Diminishing Nursing Faculty

The growing deficit of full-time master's and doctorally prepared nursing faculty is intensifying the overall nursing shortage. This lack of faculty is contributing to the current nursing shortage by curtailing the number of students admitted to nursing programs (AACN, 2003). A survey conducted by the American Association of Colleges of Nursing (AACN) found that 5,283 qualified applications to baccalaureate, master's, and doctoral programs were rejected in 2002-2003 and that 41.7 percent of responding schools cited insufficient faculty as a reason for not accepting all qualified applicants (Berlin et al., 2003).

In 2001-2002, of the 457 doctoral graduates 28.6 percent reported employment commitments in settings other than schools of nursing (Berlin et al., 2003). Data collected by the National Sample Survey of Registered Nurses for the years 1992, 1996, and 2000 showed a steady decline in the proportion of nurses with nursing doctorates who were employed in schools of nursing with baccalaureate and higher degrees, from 68 percent in 1992 to 49 percent in 2000 (Division of Nursing, 2001). Of those that complete graduate education, salary is an influential factor in employment decisions. The decisions of master's-prepared nurses to return to doctoral study rest on calculations about whether it profits them to enter academia and seek doctoral study when they could earn higher salaries in nonacademic master's-level positions (AACN, 2003).

The pipeline from enrollees to graduates of doctoral programs in nursing schools is diminishing; in the fall of 2002 the 81 research-focused doctoral programs in nursing reported 3,168 enrollees and 457 graduates. Schools are not producing more graduates even though the number of doctoral programs increased from 54 in 1992 to 83 (including two clinic-focused programs) in 2002. Trends in master's education should be of concern; in a five-year cohort of 289 schools reporting data each year, enrollments steadily declined from 1998 to 2001, followed by an increase of 898 students in 2002. Analysis indicated an average decrease of 110 students per year, despite this increase (Berlin et al., 2003). Master's graduates are the source of future doctoral students and are a significant portion of current and future faculty (AACN, 2003).

The Aging Workforce

The growth in the number of R.N.s is being limited by declining enrollments in nursing schools and the aging of the R.N. workforce (IOM,

2004b). Since 1973 the percentage of college freshmen indicating nursing as a top career choice has dropped by 40 percent (IOM, 2004b). In 1983 the average age of the R.N. workforce was 37.4 (Buerhaus et al., 2000), but this average increased to 45.2 years in 2000 (Spratley et al., 2000). The seventh national sample survey of registered nurses in the United States revealed that in the two decades from 1980 to 2000, the percentage of nurses younger than 40 dropped from 52.9 percent in 1980 to 31.7 percent in 2000 (NRC, 2000; HRSA, 2003). Furthermore, the percentage of R.N.s younger than 30 dropped from 26 percent in 1980 to less than 10 percent in 2000. In 2000 four times as many 40-year-olds as 20-year-olds were nurses (IOM, 2004b). The average age of R.N.s is projected to increase and peak at 45.5 years in 2010 (Buerhaus et al., 2000). By contrast, the Department of Labor has forecast that the average age of the overall labor force will reach only 40.7 years by 2008 (Bureau of Labor Statistics, 1999). The projections by the Health Resources and Services Administration (HRSA) for the supply and demand of R.N.s between 2000 and 2020 predict a need for 750,000 more R.N.s than will be available (HRSA, 2002).

If R.N.s are in short supply, doctorally trained nurses are growing particularly scarce. On average, they complete their degrees much later in life than do Ph.D.'s in other fields. Often this delay results from pursuing a Ph.D. part-time. Even those receiving National Research Service Award funds, which require full-time study, are generally over 40 by the time they finish their studies (Gordon et al., 2003; HRSA, 2003; McGivern, 2003). In 1999, of the 365 recipients of nursing doctoral degrees who reported their age, the median age was 46.2 years. Almost half of all graduates were between 45 and 54 years; twelve percent were older than 55; and 25 percent were younger than 35 (AACN, 2003). By comparison, the median age of all research doctoral awards in the United States was 33.7 years in 1999 (National Opinion Research Center, 2001). The median time that elapsed between entry into a master's program to completion of the doctorate in nursing was almost twice that of other fields, 15.9 and 8.5 years, respectively (National Opinion Research Center, 2001). The advanced age of nursing Ph.D.s stems in part from the norms of the profession, which encourages its members to acquire considerable professional experience before seeking research training. Although this practice ensures professional expertise, later research training inevitably limits the length of an individual's research career. The median age of nursing school faculty is now 50, and many nursing school deans report concerns about their abilities to replace retiring faculty (Gordon et al., 2003; HRSA, 2003; McGivern, 2003).

Ph.D.s in Clinical Research

Ph.D.-trained scientists are now fulfilling a wide set of roles in medical education and research (Miller, 2001). As such, they are shouldering a significant portion of work within the clinical research enterprise. Ph.D.-trained faculty in clinical departments contribute substantially to teaching, especially during the first two years of medical school. Their contribution to the research conducted by the clinical departments to which they belong has also become significant. Moreover, through their collaborative and principal investigator efforts, they fulfill important mentoring roles for clinical research trainees. Undoubtedly these scientists will continue to be part of clinical departments, especially at research-intensive academic health centers. In this regard, leaders in the field have underscored the need for alternative career tracks for these faculty members, as well as greater job stability to compensate them for their contributions. Clinical research teams of the future will likely continue to draw on Ph.D.-trained researchers. The evaluation of clinical research training programs has been proposed in previous reports (IOM, 1994; NIH, 1997; Wolf, 2002).

Other Investigators

Of the approximate 4,000 dental graduates each year in the United States, 1.5 percent express interest in academia and less than 0.2 percent are interested in research (Juliano and Oxford, 2001; Stashenko et al., 2002). U.S. dental schools report challenges in filling academic positions (Stashenko et al., 2002). Efforts to increase interest in dental research early on as well as dental research training programs are needed.

The shortage of pharmacists is a challenge as well for the clinical research enterprise. Federal pharmacy positions have experienced dramatically rising vacancy rates in recent years, reaching 11 percent in the U.S. Public Health Service and 15-18 percent in the armed forces (HRSA, 2000). In the late 1990s the number of pharmacy graduates declined, with a corresponding decline in the number of applications to pharmacy schools; in 1999 the number of applications was 33 percent lower than it had been in 1994, which was the past decade's high point (HRSA, 2000). The demand for pharmaceutical care services has grown more rapidly in the past decade. Two major components of the increase in demand have become increasingly apparent, demonstrated by (1) the increased vacancy rates and difficulties in hiring, and (2) the demand for pharmaceutical care services resulting from the increases in prescription drug volume and the expanded responsibilities and roles of today's pharmacist (HRSA, 2000).

FUTURE NEEDS

A new model for training clinical investigators is emerging; formal clinical research training programs are replacing on-the-job training (Wolf, 2002; Zerhouni, 2003). Clinical research trainees must acquire specific expertise in study design, epidemiology, and biostatistics, to name a few areas (Wolf, 2002). They must also learn when and how to most effectively apply state-of-the-art techniques of clinical research, such as genomics and proteomics. Furthermore, they must be trained in those issues that pertain specifically to research involving human subjects, such as the principles of informed consent and human safety protection.

The complex nature of clinical research requires a team approach in which investigators interact with their team members across disciplines and geographical locations. Because of the past and present state of the clinical research workforce, proactive steps are needed to develop new paradigms for a diverse and capable clinical research workforce that meets the needs of twenty-first-century medicine. As the complexity and volume of research in the life sciences have increased, groups of various investigators have tended to pool together to tackle complex problems (Drenth, 1998; Cheung et al., 2001; HRSA, 2002; Collins et al., 2003). The future will see the need for more such research teams and thus the need to promote all the potential players of the clinical research enterprise.

3

The Status of Women and Underrepresented Minorities and Programs of Support

A clinical research career may pose special challenges for women and minorities. This chapter focuses on the status of women and minorities in academic research careers, from students to faculty. Some programs that provide support and guidance to advance women and minorities in research careers are highlighted.

WOMEN FACULTY

The recruitment, retention, and advancement of women in academic medicine are critical issues for the clinical research community.

Bumpy Career Paths

Data from the Association of American Medical Colleges (AAMC), which collects and publishes data on the status of women at all levels along the medical career path, indicates that although the numbers of women applying to, enrolling in, and matriculating from medical schools continue to rise, advancement along the faculty career path has been much slower than anticipated (AAMC, 2004b). Since the establishment of the Office of Research on Women's Health at the National Institutes of Health (NIH), researchers have continually examined the progression of women in basic biomedical and clinical research careers (NIH, 1992, 1999; NRC, 2004). A workshop that focused on women in clinical research careers suggested that the varying career paths, debt burdens, and need to balance family and career had differentially acute impacts on women (NRC, 2004).

Women constitute 29 percent of the faculty of basic science departments and 30 percent of the faculty of clinical departments (Barzansky and Etzel, 2002). The data indicate that women do not advance along the academic career path at the same rate as men (see Table 3-1) (AAMC, 2003).

In a landmark study conducted in 1999, female graduates of medical schools were found more likely than male graduates to pursue an academic career, but the numbers of women advancing to associate and full professor rank were lower than expected for both tenure and nontenure tracks (Nonnemaker, 2000). The study found that 25 percent fewer women than expected rose to the rank of associate professor and 43 percent fewer women than expected rose to the rank of full professor. The influx of women into academic health professions over the past three decades has not been accompanied by equality for male and female faculty in rank attainment, leadership roles, salaries, or treatment by colleagues and superiors. An examination of one academic institution indicated substantial gender differences in the rewards and opportunities offered to men and women. There were also significant gender disparities in salary (Wright et al., 2003). After adjusting for rank, track, degree, specialty, years in rank, and administrative positions, researchers found that the women in the institution earned 11 percent less than men. In general, however, the women were as productive as the men based on both publications and clinical revenues,

TABLE 3-1 Distribution of Full-Time U.S. Medical School Faculty by Sex and Rank, 2003

Rank	Male	Female	Total
Professor	21,947	3,652	25,599
Associate professor	16,884	6,025	22,909
Assistant professor	27,072	16,186	43,258
Instructor	6,113	5,787	11,900
Other	1,070	940	2,010
Total	73,086	32,590	105,676

NOTE: The table excludes 231 faculty with missing sex data.
SOURCE: Faculty Roster, Association of American Medical Colleges, 2003.

despite having less research space and less influence in their departments. Although the women aspired to leadership positions and felt they had leadership skills, few had been asked to lead. Also noted in the study, one-third of the women reported experiencing discrimination.

A study conducted in 2001 by Morahan et al. found that seven diverse medical schools that had a U.S. Department of Health and Human Services' Office Center of Excellence in Women's Health had documented large increases in the numbers of women in senior faculty ranks. The number of senior women faculty at one institution increased from 60 to 104 (58 percent) during 1994-1999, compared with an increase from 489 to 542 (11 percent) in the number of senior men faculty. The number of tenured women faculty went up from 51 to 77 (66 percent), compared with an increase from 454 to 475 (5 percent) in the number of men.

Nationally, however, women are still underrepresented in the senior faculty ranks and administrative positions in U.S. medical schools (Morahan et al., 2001). A cross-sectional survey of all salaried physicians in 126 academic departments of pediatrics in the United States revealed that the rank of associate professor or higher was achieved by significantly more men than women. Women in the lower ranks were not as productive academically and spent a lot more time in teaching and patient care than did men in those ranks (Kaplan et al., 1996).

A study that quantified the magnitude of difference in the career advancement of clinician-educator faculty versus research faculty revealed that even after adjusting for other factors, men were almost three times more likely to be at a higher rank in academic medicine than women (Thomas et al., 2004). A multi-institutional study found that women faculty had less institutional support (e.g., research funding, secretarial support) and low satisfaction with career progression (Carr et al., 1998). Compared with men in terms of leadership and national recognition, women faculty were assigned a lower value. In addition, women faculty had the poorest understanding of promotion criteria and the least amount of time available for scholarly activities (Buckley et al., 2000b).

Special Challenges for Women Faculty

Although a career in clinical research is challenging for anyone, women must deal with considerations that make their entry more challenging. A major difficulty is timing, because the years of most productive career building coincide with the childbearing years. Several studies have noted the difficulties that women face in academic settings, including the challenge of combining family responsibilities with academic success (NIH, 1998; Bickel, 2001; NRC, 2001; Yedidia and Bickel, 2001; Bickel et al., 2002; Guelich et al., 2002; Pendharkar, 2003; Wright et al., 2003).

Because women tend to carry more family responsibilities than men (Thomas et al., 2004), women are more likely to seek flexible job arrangements to accommodate their families. A survey of women faculty at one institution found a flexible work environment without negative consequences for women with young children to be the highest ranked need (McGuire et al., 2004). A study of institutional policies on tenure, promotions, and benefits for part-time faculty at U.S. medical schools demonstrated that women were more likely to choose part-time work to balance employment with family responsibilities, whereas men were more likely to choose part-time work as a way to balance competing professional options. The advantages of part-time status differed between men and women; women cited increased involvement with children, more time for family, balance in life or work, and additional time for personal pursuits or development, whereas men cited satisfaction from teaching and an ability to keep up with the developments in the field, greater involvement in academic pursuits, and increased income from participation in other pursuits (Socolar et al., 2000). Although the American College of Physicians recommended that all medical schools develop flexibility in tenure and promotion procedures in order to help faculty accommodate personal and family responsibilities while continuing academic work—and called specifically for part-time work for faculty (ACP, 1991)—the study found that the majority of medical schools do not have policies that foster tenure for part-time faculty, although many offer a variety of benefits and may allow for promotion.

A recent study by the Office on Women's Health of the reentry of professionals into health professions found that, because of child care obligations, 90 percent of women physicians made career changes. Women were twice as likely as men to suspend their careers to yield to a spouse or a partner (Mark and Gupta, 2002).

In a recent survey of department chairs several respondents indicated that time constraints, coupled with the inflexibility of academic routines and promotion processes, were inhibitors of the advancement of women (Yedidia and Bickel, 2001).

WOMEN MEDICAL SCHOOL STUDENTS

The number of women enrolling in and matriculating from U.S. medical schools continues to grow. In 2003, for the first time, more women than men applied to medical school. Ninety-six percent of the increase in applicants in 2003 (over 1,100) was attributable to women. That same year women made up 50 percent of first-year medical students and 46 percent of medical graduates (AAMC, 2004b). The rate of growth of women students indicates that women will constitute the majority of students and graduates in the next decade.

Women medical students face some of the same issues, along a continuum, as women faculty and residents as well as some of the same concerns about debt burdens facing underrepresented minority students. These concerns include decisions about childbearing and family responsibilities and their longer career paths. Women's interest in research (though not academic) careers has declined at a slightly higher rate than that of men (Bickel, 2004).

UNDERREPRESENTED MINORITY FACULTY

Blacks, Hispanics, American Indians, and Alaskan Natives remain underrepresented in science and academia, despite significant efforts in recent years to increase diversity in these fields (Crowley et al., 2004). The low number of minority medical school faculty members, especially at the tenured faculty level, reduces the pool of available candidates for physician clinical research investigators (NIH, 2002b). In 2002, 77.0 percent of U.S. medical school faculty members were white, 11.5 percent Asian, 3.8 percent Hispanic, 3.0 percent black, 4.6 percent other, and 0.1 percent American Indian (AAMC, 2002).

Distribution of Underrepresented Faculty

Only 4.2 percent of U.S. medical schools have underrepresented minorities in their faculty; indeed, faculty at six schools alone account for approximately 20 percent of underrepresented faculty in the United States. When these six schools are excluded, the underrepresented faculty at other U.S. medical schools drops to 3.5 percent (AAMC, 2002). Only at seven institutions do underrepresented minorities constitute more than 10 percent of the faculty. Overall, underrepresented minorities represent 4.5 percent of the clinical faculty at all U.S. medical schools. Between 1980 and 2000 the number of underrepresented minority faculty increased 279 percent (see Figure 3-1). Most underrepresented faculty are in the assistant professor and instructor ranks (see Figure 3-2).

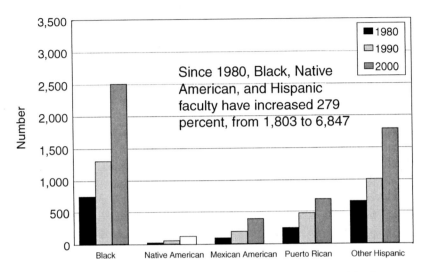

FIGURE 3-1 Black, American Indian, and Hispanic U.S. medical school faculty, 1980-2000.
SOURCE: AAMC Faculty Roster.

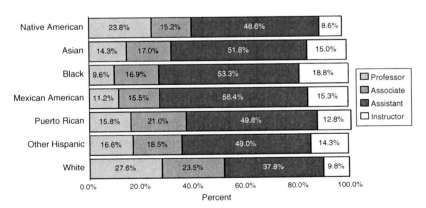

FIGURE 3-2 Medical school faculty by race and ethnicity, 2002.
SOURCE: AAMC Faculty Roster System, December 31, 2002.

Special Challenges for Minority Faculty

Underrepresented clinical research scientists face many of the same challenges confronting well-represented ones, including dealing with a lack of support from their institutions or departments, finding mentors who can alleviate the difficulties of entry into the established research infrastructure, and balancing professional interests with the realities of having to repay educational loans (Lee et al., 2001; NIH, 2002a). It is not unusual for underrepresented clinical research scientists to face institutional biases in support because of the lack of support systems—they are often the only, or among only a few, underrepresented minority members in an academic health center or research institution. Because of underrepresentation within departments and institutions, such minority faculty are often asked to represent the department or institution and to serve on multiple committees, creating an additional time burden.

Gartland et al. (2003) compared the satisfaction of black physicians and the satisfaction of white physicians with their medical schools, their medical careers, their professional and research activities, and achievements. They found that black physicians were more dissatisfied than white physicians with the social environment of medical school. The small number of underrepresented faculty reduces the likelihood that underrepresented senior faculty can mentor underrepresented junior faculty. The situation also affects the medical students. There is less likelihood of underrepresented students finding an underrepresented faculty member to serve as an advocate and provide survival strategies. For those underrepresented faculty who are on staff, serving as a role model should be added to their roles as advocates and providers of survival strategies.

UNDERREPRESENTED MINORITY STUDENTS IN MEDICAL SCHOOLS

Of the 17,592 students entering the 2002 U.S. medical school class 2,013 students identified themselves as underrepresented minorities. The acceptance rates for underrepresented minority students were slightly below those of Asians and whites, and distinct Hispanic groups had higher acceptance rates than "other Hispanic" groups (see Table 3-2). The overall number of minority medical school graduates has increased during the last 10 years, but the denominator has increased as well (see Figure 3-3). Thus, the proportion remains virtually unchanged.

TABLE 3-2 Hispanic Ethnicity and Non-Hispanic Race Medical School Applicants by Acceptance Status, 2002 and 2003

		2002		
Applicants		Applicants	First-Time Applicants	Acceptees
Hispanic	Mexican American	738	539	379
	Puerto Rican	634	505	329
	Cuban	142	106	70
	Other Hispanic	804	551	351
	Multiple Hispanic	123	87	64
	Subtotal	2,441	1,788	1,193
Non-Hispanic	Black	2,611	1,790	1,178
	Asian	5,949	4,341	3,219
	American Indian (including Alaskan Natives)	112	80	56
	Native Hawaiian, Other Pacific Islanders	36	26	12
	White	19,454	14,553	10,649
	Other	526	367	225
	Unknown	282	261	190
	Multiple race	1,235	916	596
	Subtotal	30,205	22,334	16,125
Non-U.S.	Foreign	943	730	274
	Unknown Citizenship	36	35	1
	Subtotal	979	765	275
Total		33,625	24,887	17,593

SOURCE: Data Warehouse, Applicant Matriculate File as of November 6, 2003, Association of American Medical Colleges.

Only about 250 more blacks received M.D. degrees in 2001 than in 1975. In 1971-1972 only 9 percent of medical students were black, American Indian, Hispanic, or Asian or Pacific Islander (Barzansky and Etzel, 2002). Unfortunately, the number of underrepresented graduates with an expressed interest in research careers is dwindling (NIH, 2002b).

Debt is a significant issue for all students, but it is a particularly daunting one for minority students. Black and Mexican American students

		2003		
Matriculants	Applicants	First-Time Applicants	Acceptees	Matriculants
355	773	576	361	336
312	522	403	291	280
64	154	112	78	70
336	878	607	353	345
63	156	115	61	58
1,130	2,483	1,813	1,144	1,089
1,125	2,740	1,905	1,123	1,060
3,030	6,152	4,566	3,196	3,065
49	85	61	38	35
12	23	15	6	5
9,972	20,231	15,466	10,749	10,122
214	599	450	223	206
184	188	170	132	121
562	1,328	961	667	626
15,148	31,346	23,594	16,134	15,240
209	957	753	261	209
1	0	0	0	0
210	957	753	261	209
16,488	34,786	26,160	17,539	16,538

have a higher level of education debt than do Asians and Puerto Ricans (AAMC, 2004a). Median indebtedness levels are slightly higher for black students and somewhat lower for Asians, Mexican Americans, and Puerto Ricans (AAMC, 2004a). The prospect of accumulating additional debt is off-putting, especially when the acquired debt is greater than the family income for a year.

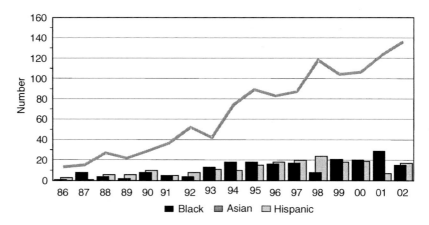

FIGURE 3-3 Black, Asian, and Hispanic M.D./Ph.D. graduates, 1986-2002.
SOURCE: Data Warehouse, Association of American Medical Colleges, 2002.

NIH PROGRAMS FOR CLINICAL RESEARCH AND MINORITY RESEARCHERS

The National Institutes of Health administer a variety of public programs that may be used to help develop minority clinical researchers. Some of these programs are directed toward developing clinical researchers generally regardless of race or ethnicity, though some grants may be administered through a minority-targeted component. Other programs are specifically targeted to the development of minority investigators, some of whom will become basic biomedical researchers and others will become clinical investigators.

Minority Research Training Programs

The NIH provides targeted programs designed to increase the participation of underrepresented minorities in biomedical, behavioral, and clinical research careers. These programs for undergraduates, graduate students, and postdoctoral fellows seek to increase the participation in these fields of individuals from historically underrepresented groups: African Americans, Hispanics, American Indians, and Pacific Islanders. While clinical training typically occurs at the graduate or postdoctoral level, undergraduate programs discussed below may train students who later choose a clinical research career. Examples of minority-targeted programs are profiled in Appendix D.

The NIH supports undergraduate education for underrepresented minorities in the biomedical and behavioral sciences most directly through three programs offered by the National Institute of General Medical Sciences (NIGMS) and the National Institute of Mental Health (NIMH). The first of these is the Bridges to the Baccalaureate (R25) program, which focuses on the preparation of students in the biomedical or behavioral sciences at two-year institutions—community or tribal colleges—in order to prepare them for transfer to a four-year institution. The other two programs are the NIGMS Minority Access to Research Careers (MARC) Undergraduate Student Training in Academic Research Program (U*STAR) (T34) and the NIMH Career Opportunities in Research Education and Training (COR) (T34). These programs focus on students in their third and fourth years of undergraduate study. All three programs are administered through institutional awards to historically black colleges and universities, Hispanic-serving institutions, or tribal colleges or universities. The programs provide students with coursework, hands-on research experience, mentoring, career counseling, and financial support.

At the graduate and postdoctoral levels NIH provides a variety of individual and institutional awards. Two National Research Service Award (NRSA) fellowship programs are targeted to minorities through the F31 mechanism: the NRSA Predoctoral Fellowship for Minority Students and the MARC Predoctoral Fellowship Program. The latter is targeted at graduates of the U*STAR program. The F31 fellowship provides an annual stipend, tuition, and fee allowance as well as an annual institutional allowance that may be used for travel to scientific meetings and for laboratory and other training expenses. The NIMH Research Grants to Increase Diversity in the Mental Health Research Arena support minority students in mental-health-related fields working on their dissertations. NRSA Institutional Training Grants (T32) and Short-Term Institutional Training Grants (T35) targeted to minority-serving institutions also seek to increase the participation of underrepresented minorities. The National Heart, Lung, and Blood Institute, for example, has utilized the T32 and T35 mechanisms to encourage the development of minorities in cardiovascular, pulmonary, hematological, and sleep disorders research fields.

The National Academies completed an assessment of NIH's minority research training programs in early 2005, and a more complete listing and description of NIH minority-targeted programs can be found in that report

(NRC, 2005b). That assessment noted that the number and percentage of minorities earning Ph.D.s in the biomedical sciences over the last decade have been relatively flat. Still, the assessment concluded that without the availability of minority-targeted programs—which provided important financial support, mentoring, coursework, and research experience—the numbers and percentages may well have declined.

Clinical Workforce Programs

NIH also administers training programs specifically designed to increase the clinical research workforce. These programs include the Institutional Research Training Grant (T32), the Short-Term Institutional Training Grants (T35), Mentored Clinical Science Development Award (K08), Mentored Clinical Science Development Program Award (K12), Mentored Patient-Oriented Research Career Development Award (K23), Midcareer Investigator Award (K24), and Clinical Research Curriculum Award (K30). Appendix E provides profiles of these programs.

Loan Repayment Programs

NIH has a variety of loan repayment programs (LRPs) to support the recruitment and retention of health professionals as clinical or pediatric investigators. Loan repayment programs have also recently been introduced to increase the clinical research workforce in general, and women and minorities have been strongly encouraged to apply. The LRPs allow repayment of up to $35,000 of the principal and interest of eligible educational loans of clinical or pediatric investigators for each year of research service, and the payment of 39 percent of the loan repayment amount per year toward federal tax liability prevention. The LRP is a contractual agreement, in which awardees agree to engage in clinical or pediatric research for a minimum of two years.

Examples of these repayment programs are the Health Disparities LRP, the Clinical Research LRP for Individuals from Disadvantaged Backgrounds, the Clinical Research LRP, the Pediatric Research LRP, and the Contraception and Infertility Research LRP. Three of these LRPs have seen an increase in applications (see Figure 3-4): the Clinical Research LRP had 1,150 applicants in FY 2003 compared with 487 in FY 2002; the Health Disparities LRP had 182 applicants in FY 2003 compared with 170in FY 2002; and the Pediatric Research LRP had 494 applicants in FY 2003 com-

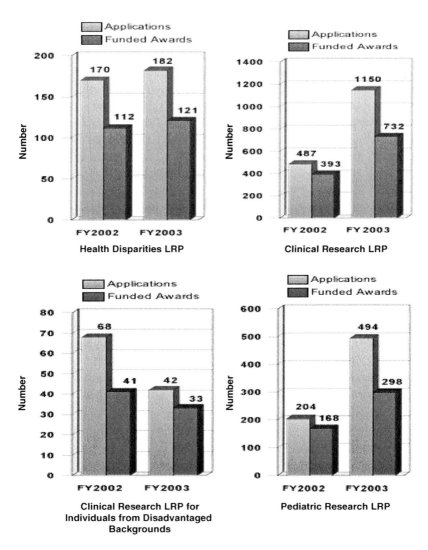

FIGURE 3-4 New applications and funded awards for four NIH loan repayment programs, FY 2002 and FY 2003.
SOURCE: National Institutes of Health, http://www.lrp.nih.gov.

TABLE 3-3 Distribution of Loan Repayment Program Applicants by Sex, FY 2003

Sex	Number of Applications	Number Funded	Success Rate (%)
Clinical Research LRP			
Unknown	33	18	55
Female	519	347	67
Male	541	362	67
Clinical Research LRP for Individuals from Disadvantaged Backgrounds			
Unknown	3	2	67
Female	26	21	81
Male	13	10	77
Health Disparities LRP			
Unknown	2	2	100
Female	112	79	71
Male	68	40	59
Pediatric Research LRP			
Unknown	15	5	33
Female	246	148	60
Male	219	146	67

SOURCE: David Conboy, associate director for policy and liaison activities, Office of Loan Repayment, National Institutes of Health.

pared with 204 in FY 2002. By contrast, the Clinical Research LRP for Individuals from Disadvantaged Backgrounds saw a decrease in applications—42 applicants in FY 2003 compared with 68 in FY 2002 (see Figure 3-4). In FY 2003, 1,883 total LRP applications were received, and 1,200 researchers received LRP contracts. In terms of gender, the Clinical Research LRP, the Clinical Research LRP for Individuals from Disadvantaged Backgrounds, and the Pediatric Research LRP all had equal or greater percentages of women funded compared with men (see Table 3-3). Total LRP contracts reached $63.3 million in FY 2003. These programs are a promising development in addressing the financial disincentives to clinical research careers.

Grant Supplements

Supplements to research grants[1] were established by NIH to address the need to increase the number of underrepresented minority scientists participating in biomedical research and the health-related sciences.

Programs to Advance Women's Research Careers

The NIH Office of Research on Women's Health, along with a number of co-sponsors, offers Building Interdisciplinary Research Careers in Women's Health (BIRCWH) Career Development Programs. These programs support the research career development of junior faculty members, known as Interdisciplinary Women's Health Research (IWHR) Scholars, who have recently completed clinical training or postdoctoral fellowships and who are commencing basic, translational, behavioral, clinical, or health services research relevant to women's health.

The programs aim to bridge advanced training with research independence, as well as to connect scientific disciplines or areas of interest, via the mentored research career development award (K12) mechanism. They will therefore increase the number and skills of investigators at awardee institutions through a mentored research and career development experience, leading to an independent interdisciplinary scientific career addressing women's health.

DEPARTMENT OF VETERANS AFFAIRS PROGRAMS

The Office of Research and Development (ORD) at the Department of Veterans Affairs (VA) has one very broad initiative aimed at developing clinical researchers and one aimed at developing a cadre of investigators devoted to increasing knowledge of racial disparities in health and health care and other issues related to the quality of care or health services across racial boundaries.

Currently 13 VA Health Services Research and Development (HSR&D) Centers of Excellence (COE) and four resource centers are located throughout the United States. Each COE develops its own research agenda, is affili-

---

[1] See http://grants.nih.gov/grants/guide/pa-files/PA-01-079.html. Date accessed November 19, 2004.

ated with a VA medical center, and collaborates with local schools of public health and universities. COE research covers an array of important healthcare topics, such as quality of care, chronic diseases, primary care, mental health, substance abuse, pain management, and outcomes research. The four resource centers provide support and information to VA researchers and healthcare managers in the special areas of management research, data and information sources within and outside the VA, health economics and cost studies, and measurement of knowledge and instruments.[2]

Diversity-Building Research Training Program

The VA's new Diversity-Building Research Training Program is aimed at proactively recruiting and retaining a diverse healthcare research team. Individuals and academic institutions that can present a unique perspective and supply important insight into relevant cultural factors that may account for health disparities among veterans and who can successfully report how their backgrounds and personal achievements can contribute to VA research are highly encouraged to apply for three new awards.[3] The Mentoring Research Enhancement Coordinating Center Award advocates institutional collaboration between the VA and institutions of higher learning, including but not limited to historically black colleges and universities, Hispanic-serving institutions, and tribal colleges and universities that are committed to achieving diversity in the biomedical sciences and that deliver encouragement, support, and guidance to students from a myriad of backgrounds and with a myriad of personal achievements. The Mentored Supplemental Award (one-on-one training) is for applied training in research on VA-funded research projects. The Mentored Early Career Enhancement Award (one-on-one training) offers an encouraging career path for mentored research in the VA.

Training Opportunities: National Networks

The VA aims to generate national networks of training opportunities with its clinical research Centers of Excellence. It is concentrating first on core funding for methodologists who will ensure advancement in tools, methods, and measures for health services research, and toward that end it

---

[2] See http://www.hsrd.research.va.gov. Date accessed November 19, 2004.
[3] Ibid.

will provide $400,000 a year to support methodologists with Ph.D.'s. The Seattle Epidemiologic Research and Information Center, VA Employee Education System, and University of Washington are supporting the six distance-learning, cyber-session courses being conducted for VA researchers, clinicians, and administrators in medical centers through the VA Knowledge Network satellite system. The six classes are:

1. Developing scientific research proposals (grant writing);
2. Applied regression analysis;
3. Advanced issues in clinical trials using the Women's Health Initiative as an example;
4. Cost and outcomes research;
5. Clinical trials; and
6. General biostatistics.

When these methodologist positions are completely filled, the VA will provide more resources and support for clinicians nationally so that they will help the VA to determine good research questions.

PRIVATE SOURCES OF FUNDING FOR CLINICAL INVESTIGATORS

Several foundations and voluntary health associations offer funding and training for clinical investigators. Notable examples include the Burroughs Wellcome Fund, Doris Duke Charitable Foundation, Howard Hughes Medical Institute, American Heart Association, American Diabetes Association, American Cancer Society, Juvenile Diabetes Research Foundation, and Robert Wood Johnson Foundation. An analysis of funding of clinical research by 11 private foundations identified a $259 million commitment from 1997 to 2001 for career development of clinical investigators, including training and research support (Nathan and Wilson, 2003). These foundations and others also offer specific programs for minority health professional education; examples are the Ford Foundation, W. K. Kellogg Foundation, and California Endowment.

Howard Hughes Medical Institute

With the support of the Howard Hughes Medical Institute-National Institutes of Health (HHMI-NIH) Medical Scholars Program, known as the Cloister Program, selected students spend the year conducting research at NIH. In a second program, the HHMI Medical Fellows Program, students spend a full year doing research at their own or at another institution.[4] Forty-two individuals are selected each year for the Cloister Program, and 60 students participate in the Medical Fellows Program. Two-thirds of the students entering these programs have completed their second year of medical school; the rest have completed their third year. HHMI provides a modest stipend (between $18,000 and $24,000) and pays for some of their supplies.

The goal of the HHMI Research Training Fellowships for Medical Students is to strengthen and expand the nation's pool of medically trained researchers. The fellowships provide funds to support fellows and cover their research- and education-related expenses. Through annual competitions HHMI provides three types of medical student fellowships under this program: (1) support for an initial year of research training, (2) continued support for research training, and (3) continued support for completion of medical studies. In 2004 HHMI awarded up to 60 fellowships to medical and dental students who show the greatest promise for future achievement in biomedical research and who have demonstrated superior scholarship as undergraduates and during their initial medical or dental school training.

Eleven percent of individuals in the Cloister Program and the Medical Fellows Program are women and minorities. The HHMI programs, with one year of investment, compete with the other M.D.-Ph.D. programs around the country for promoting participation in research. Of the students from the 1985 and 1986 fellowship years who are still conducting research, virtually all are engaged in translational or clinical research.

Robert Wood Johnson Foundation

Within the Robert Wood Johnson Foundation (RWJF) Program are two programs for clinical researchers—the Clinical Scholars Program and the Generalist Physician Faculty Scholars.[5] The Clinical Scholars Program

---

[4]See http://www.hhmi.org/research/cloister and http://www.hhmi.org/grants/funding/comp_annc/2004_med_pa.pdf. Date accessed October 15, 2004.

[5]More information can be found on all three programs at http://www.rwjf.org/index.jsp, http://rwjcsp.stanford.edu/index.html, and http://www.gpscholar.uthscsa.edu/gpscholar/FacultyScholars/about.html. Date accessed November 22, 2004.

is designed to augment clinical training by providing the new skills and perspectives necessary to achieving leadership positions both inside and outside academia in the twenty-first century. The program stresses training in the quantitative and qualitative sciences that underlie health services research and are important to improving health and medical care systems. In the program's newest iteration there will be an additional emphasis on community-based research and leadership training. The RWJF's Generalist Physician Faculty Scholars Program awards four-year career development grants to outstanding junior faculty at U.S. medical schools in family practice, general internal medicine, and general pediatrics. This program is intended to strengthen generalist physician faculty in the nation's medical schools by improving their research capacity while maintaining their clinical and teaching competencies.

In 1972 the first activities of the newly established RWJF were scholarship and loan programs for women, minorities, and people interested in the medical professions from rural areas. RWJF engaged in the national medical fellowships and encouraged the University for Medicine and Dentistry of New Jersey (UMDNJ)[6] to launch a summer enrichment program for minority students entering medical or dental school. Since then all these programs have grown significantly. Founded in 1962, the Robert Wood Johnson Medical School is one of eight schools of the UMDNJ.

In 2003 the RWJF and Kellogg Foundation, working with the Association of American Medical Colleges, formed the Health Professionals Partnership Initiative (HPPI).[7] They created 26 partnerships, 5 in the area of public health, with the goal of leading and helping medical and professional schools to create an environment in which they work in partnership with communities and high schools to enable more students to go into the health professions.

The Minority Medical Faculty Development Program[8] provides support for minority medical faculty who spend up to 70 percent of their time in research. Although this program initially focused on basic research,

---

[6]See http://rwjms.umd.nj.edu. Date accessed October 14, 2004.

[7]See http://www.rwjf.org/publications/publicationsPdfs/health-prof-partnership.pdf. Date accessed October 27, 2004.

[8]See http://www.mmfdp.org/about.htm. Date accessed October 27, 2004.

in the last decade RWJF has shifted toward clinical research. The Minority Medical Faculty Development Program seeks to increase the number of minority faculty who achieve senior rank in academic medicine and who will encourage and foster the development of succeeding classes of minority physicians. A key component of this program is mentorship, which is also one reason for its success. More than 100 fellows have completed all four years of the program. Of these, more than 80 percent are still in academic medicine.

FUTURE DIRECTIONS

To achieve a robust, diverse clinical research workforce, systemic change in approaches to education, training, and career development is needed in the culture of academic health centers. (See Summary in Box 3-1.) Diversity should be incentivized and institutionalized into the mission, operations, and reward structure of academic health centers. Review and evaluation of current strategies to recruit, retain, and advance women and minorities are needed to identify successes, which could then be disseminated and adopted more widely. Programs that have been shown to work could be expanded and established at other institutions. If a critical mass of women and underrepresented minorities can be achieved, diversity may become self-sustaining.

As our knowledge of human health increases, so do the number of research questions about human disease and treatment. Diversity of views can bring diversity of approaches to research problems, issues, and topics, which can contribute to the richness of our understanding.

**BOX 3-1
Summary**

Approaches to Increasing Diversity
- Recruit more underrepresented minorities for medical school;
- Encourage more women and underrepresented minorities to consider careers that include research; and
- Promote underrepresented medical students coming out of programs such as NIH's Minority Access to Research Careers (MARC) and Minority Biomedical Research Support (MBRS) Programs and the HHMI Exceptional Research Opportunities Program (EXROP).

U.S. Medical School Faculty, 2002
- As of December 31, 2002, there were 98,802 full-time faculty at 126 medical schools;
- Nearly 84 percent were in clinical departments;
- Women represented 29 percent of all faculty; and
- Underrepresented minorities represented 7 percent of all faculty.

Challenges for Faculty
- Isolation, debt, time, lack of mentorship, generating clinical revenue;
- As medical students, there is a cohort effect, that is, groups of students spend time together in classes and other efforts (as faculty, individuals are part of a larger organizational structure with multiple academic departments, sections within departments, and various clinical locations—a person of color is often "the only one");
- Being the female and/or minority voice on committees as well as role model and mentor;
- Being the "face" to the community; and
- Coping with additional expectations from family and "community."

Resources for Minority Faculty Development
- RWJF Minority Faculty Development Program;
- "K" Series Awards from NIH and the Agency for Healthcare Research and Quality (AHRQ); and
- National Institute of General Medical Sciences, Minority Biomedical Research Support, and EXPORT and EXCEED Grants from NIH and AHRQ.

continued

BOX 3-1 Continued

Minority Supplements
- Health Resources and Services Administration (HRSA) Centers of Excellence; and
- HRSA Minority Faculty Fellowship Program.

Future Directions
- What degree of cultural change is needed in academic health centers and disciplinary societies?
- What methods of recruitment and retention are effective for increasing diversity?
- When do you have mission imperative for diversity? and
- Need for critical mass.

4

Status and Future Role of Academic Nursing in Clinical Research

Nurses with doctoral training are needed not only to train the nursing workforce but also to conduct research and oversee research training. Yet the number of doctorally trained nurses is insufficient to meet the demand in academic and clinical settings. The lack of doctorally trained nurses to serve as faculty is a significant constraint to training nurses for practice as well as for future faculty.

The challenges facing nurse-scientists are quite daunting. In academic and clinical institutions, there is a lack of nurse mentors with a career commitment to clinical research. The creation of research-intensive environments that foster the development of students is necessary to both attract nurses into research and support their development. Nurses interested in research who work in low-intensity research environments without adequate mentors or role models feel isolated and without the necessary support to begin a clinical research career (Reame, 2003).

THE ADVANCING AGE OF NURSING FACULTY

In 2000 nearly 6,000 qualified applicants were not admitted into nursing programs, despite the shortage of nurses. In more than a third of the cases the nonadmittance stemmed from a shortage of nursing faculty. In 2003, 11,000 applicants were turned away.[1] In the next few years almost

---

[1] American Association of Colleges of Nursing at http://www.aacn.nche.edu. Date accessed November 22, 2004.

34 percent of the nursing faculty is expected to retire, exacerbating the current situation. The maturing R.N. workforce is a product of two phenomena: (1) a shrinking pool of young nurses entering the R.N. population and (2) large cohorts of the R.N. population moving into their 50s and 60s (Spratley et al., 2000).

The average age of nursing faculty members is over 50, most likely a function of the discipline's conventional late entry to doctoral study. The advanced age of nursing Ph.D.s may stem from the norms of the profession, which encourages its members to acquire considerable professional experience before seeking research training (NRC, 2000).

The average age of nurses upon completion of the doctorate is 46 years, well beyond that of other disciplines where the average age is 33 years. Those receiving National Research Service Award (NRSA) funds, which demand full-time study, are generally over 40 by the time they complete their studies (NRC, 2000). About 49 percent of all nurse-Ph.D. graduates enter the service sector rather than academia.

The Nursing Pipeline

There is an urgent need for enhanced recruitment of men and women into graduate and nursing education programs. In March 2000, R.N.s enrolled in formal education programs leading to a nursing or nursing-related degree represented only 6.7 percent of all the country's R.N.s, or 180,765 of the 2,696,540 population (Spratley et al., 2000). Enrollees tended to be part-time students (76 percent) and to be employed full-time in nursing (72 percent). Of the 180,765 nurses pursuing formal education, about 53 percent were enrolled in programs leading to a baccalaureate degree, 36.4 percent in programs leading to a master's degree, and less than 4 percent in doctoral programs.

All baccalaureate programs in nursing education have built-in components: basic research methods, statistics, and research utilization. The overwhelming majority of master's programs have a research obligation. Often this obligation takes the form of a requirement to carry out an evaluation study while receiving clinical experience or perhaps a requirement to do a secondary analysis of an existing clinical database. Doctoral training for nurses is by nature research intensive, just as it is in other disciplines. Many postdoctoral programs in nursing expect fellows to submit an individual grant proposal for external funding (e.g., NRSA) by the end of the fellowship period (McBride, 2003).

In general, students enter academic nursing programs to prepare themselves as clinicians, not researchers. Students and future clinical researchers often are not aware of the possibility of becoming clinical researchers, have incorrect assumptions about research, or believe that research would simply not be a good career match for them (Woods, 2003).[2]

Clinical Research as a Career

One of the principal pushes in the nursing field is to encourage research as a career track; B.S.N.-Ph.D. and fast-track programs are the most common mechanisms. Development of an honors program at the B.S.N. level is a positive step toward this goal. A small number of institutions offer undergraduate and graduate education and postdoctoral training in an accelerated manner and provide mentoring throughout the education (McGivern, 2003). Nursing is a field populated largely by women; multiple relocations of families for graduate, postdoctoral, and finally permanent faculty positions may not be a possibility. Another problem in nursing is that the shortage is so great that every faculty member is expected to educate more people to replenish the workforce rather than build the science (McBride, 2003).

PREPARING A DIVERSE AND REPRESENTATIVE CLINICAL RESEARCH WORKFORCE

The scientific community should encourage children's exposure to the nursing field as early as elementary and middle school to prepare a diverse and representative clinical research workforce. Graduates from baccalaureate, master's, and doctoral programs in nursing demonstrate a lack of racial and ethnic diversity (see Table 4-1). A more representative workforce will require continuing and seamless opportunities to nurture interests in clinical research careers. There is a multiplicity of programs—some that involve children in grade school, some that engage middle schoolers, and some that work with high school and college students—but they are not always coordinated.[3]

Priority should be given to retaining, not just recruiting, a representative and diverse workforce. Changing the cultural demographics of nurse

---

[2]Nancy Fugate Woods, R.N., Ph.D., Workshop Presentation: 2003.
[3]Ibid.

TABLE 4-1 Race and Ethnicity of Graduates from Baccalaureate, Master's, and Doctoral Programs in Nursing, 1999-2002

	1999-2000		2000-2001		2001-2002	
	Number	Percent	Number	Percent	Number	Percent
Total baccalaureate enrollment (generic [basic] and R.N. to baccalaureate)						
White	26,063	81.2	24,405	79.5	25,031	78.5
Black or African American	3,125	9.7	3,241	10.6	3,448	10.8
Hispanic or Latino	1,383	4.3	1,559	5.1	1,696	5.3
Asian, Native Hawaiian, Other Pacific Islander	1,332	4.1	1,293	4.2	1,480	4.6
American Indian or Alaskan Native	210	0.7	192	0.6	227	0.7
Total	32,113		30,690		31,882	
Total minority	6,050	8.8	6,285	20.5	6,851	21.5
Master's enrollment						
White	8,117	86.3	7,781	83.7	7,306	83.0
Black or African American	565	6.0	696	7.5	690	7.8
Hispanic or Latino	275	2.9	366	3.9	349	4.0
Asian, Native Hawaiian, Other Pacific Islander	400	4.3	400	4.3	397	4.5
American Indian or Alaskan Native	52	0.6	58	0.6	61	0.7
Total	9,409		9,301		8,803	
Total minority	1,292	13.7	1,520	16.3	1,497	17.0
Doctoral enrollment						
White	346	89.4	286	85.1	374	91.0
Black or African American	21	5.4	20	6.0	21	5.1
Hispanic or Latino	3	0.8	8	2.4	6	1.5
Asian, Native Hawaiian, Other Pacific Islander	16	4.1	19	5.7	10	2.4
American Indian or Alaskan Native	1	0.3	3	0.9	0	0.0
Total	387		336		411	
Total minority	41	10.6	50	14.9	37	9.0

SOURCE: AACN (2002a).

clinical researchers will require seeking initiatives that lead to both the graduation and retention of minority students (Nugent et al., 2004). One way to achieve this goal is by building learning communities of mentors who are working clinical scientists and students who represent all levels of the university curricula, bridging the disciplinary boundaries. According to the National Learning Communities Project, "In higher education, curricular learning communities are classes that are linked or clustered during an academic term, often around an interdisciplinary theme, and enroll a common cohort of students. A variety of approaches are used to build these learning communities, with all intended to restructure the students' time, credit, and learning experiences to build community among students, between students and their teachers, and among faculty members and disciplines."[4] Nurse educators are valuing diversity and cultural competence with the growing diversity of the American population (Christman, 1998).

Retaining a representative and diverse workforce can also be achieved by helping students find meaning in the work of clinical researchers. Students can understand what a study or particular health problem may mean to them personally, to their culture, or to their ethnic group. Students can see the relevance of the issue to them and to their community and can understand who will benefit from the work. Nurses can create some learning opportunities that help students address these questions as part of their training (Woods, 2003).

NATIONAL INSTITUTE OF NURSING RESEARCH

The National Institute of Nursing Research (NINR), established in 1985 as the National Center for Nursing Research at the National Institutes of Health (NIH), provides funds for training nurse-researchers and sets a national nursing research agenda.

The National Institute of Nursing Research[5] devotes about 8 percent of its budget to training, which is more than twice the average across NIH. NINR's budget support for training reflects a commitment to developing the next generation of researchers.

---

[4]National Learning Communities Project at http://www.pewundergradforum.org/project%20washington%20center.html. Date accessed December 6, 2004. See also http://learningcommons.evergreen.edu/02_nlcp_entry.asp. Date accessed November 16, 2004.

[5]National Institute of Nursing Research at http://www.nih.gov/ninr/index.html. Date accessed November 16, 2004.

In terms of success across NIH, schools of nursing, with few exceptions, are funded by all NIH institutes and centers. The NINR collaborates with other institutes and centers in many areas of shared interest, including joint funding of research project grants and requests for applications (RFAs). Collaboration extends to other agencies within the Department of Health and Human Services and beyond, including the Health Resources and Services Administration, Agency for Healthcare Research and Quality, and the Centers for Disease Control and Prevention.

As part of the NIH roadmap initiative, NINR created five expert panels to look at the future needs of nursing research. NINR's scientific goals for 2000-2004 are to (1) identify and support research opportunities that will achieve scientific distinction and produce significant contributions to health; (2) identify and support future areas of opportunity to advance research on high-quality, cost-effective care and to contribute to the scientific base for nursing practice; (3) communicate and disseminate research findings resulting from NINR-funded research; and (4) enhance the development of nurse-researchers through training and career development opportunities.

The NINR supports the research training of about 200 predoctoral students and about 70 postdoctoral fellows a year. In FY 2004 about 2,420 trainees were participating in T32 training grants in schools of nursing across the United States.[6]

The NINR supports developmental centers and institutional training awards, as well as 10 P30 grants, 8 of which focus on health disparities in minority populations. Approximately 20 percent of the NINR budget is directed toward research and training that has specific objectives related to minority health and the broader area of health disparities. NINR collaborates with historically black colleges and universities, especially those few that provide nursing education. In the broader area of health disparities NINR plans to undertake the following in the future: (1) the institute will continue to provide links between NINR-funded investigators and minority researchers who are interested in participating in large multicenter studies; (2) specific RFAs will be issued to target minority researchers and infrastructure development to support research on health disparities; and (3) the Research Supplement for Underrepresented Minority (RSUM) mechanism will continue to target minority students and faculty early in their nursing

---

[6]CRISP database at http://crisp.cit.nih.gov. Date accessed October 19, 2004.

careers to stimulate their interest in research. In addition to the developmental centers and institutional training awards, the NINR intramural program offers an intensive summer genetics institute. The institute also has developed online information for junior investigators preparing to launch their independent careers.

The NINR and the National Center for Minority Health and Health Disparities are funding exploratory centers across the country called Nursing Partnership Centers on Health Disparities (P20). These 17 centers encompass traditionally black colleges and universities and institutions serving Hispanic and American Indian students.

The centers have a two-pronged approach: (1) to boost the numbers of minority researchers and (2) to improve the quality of minority health research itself. To achieve these goals, they are funding pilot studies to entice people to enter a research career track.

FUTURE NEEDS AT THE INTERFACE OF NURSING AND CLINICAL RESEARCH

For the new paradigms in clinical research training, interdisciplinary exposure is the foundation for team science; for example, the Human Genome Project has already altered the future landscape of nursing—the underlying genetic foundation is known for diagnosis and the treatment of disease, affecting all of medicine and nursing (Horner et al., 2004). Nursing science is beginning to utilize genetic principles in research design and methodologies. Collaboration among nursing researchers and researchers in related disciplines is important for successful integration of genetic concepts into nursing science. In order to participate in the knowledge becoming available about the connections between genetics, health, and nursing, nurses must grasp genetic concepts (Williams et al., 2004). One study, which examined several surveys, found a near absence of genetics curricula in nursing schools. To address the lack of genetics contents in nursing curricula, the Genetics Program for Nursing Faculty (GPNF) was created, and it led to the formulation of a Genetics Curriculum Checklist to consolidate genetics material into curricula (Hetteberg and Prows, 2004).

Challenges also lay ahead in dealing with the nursing shortage. One workshop participant finds that a high percentage of the students who enter his clinical research administration program are nurses who either are tired of the patient care component of nursing or have left the profession for something different. The retention of nurses is an issue related to the nursing shortage, and clinical research is a tremendous draw, bringing nurses who have left back into the profession.

Another workshop participant added that even when nurses become coordinators of clinical research activities, they are perceived as having left the nursing field. Instead, clinical research needs to be seen as a part of the range of what constitutes nursing. Much of what must be done to engage people in a study, keep them enrolled, and work with them over time involves the basic skills that are part of nursing education and preparation. The percentage of nurses who do not stay in clinical nursing for longer than two years is very high. Retention is as critical an issue as recruitment into the field.

This workshop participant also observed that although there is a worldwide nursing shortage, some Asian entrants come into American doctoral programs because they are attracted to the American model of nursing education. The Asian entrants want to develop research-intensive programs, and many of them, depending on their country and the year, have full-time funding from their governments. Indeed, in a given year the strongest applicants to nursing programs have been from other countries, because they have had government funding to support full-time study. American-born individuals have not had the same kinds of resource options for doctoral study. Moreover, international students, particularly for graduate studies, do not necessarily remain in the United States; many of them come with the expectation that the support is contingent on returning to their country for at least two years. The proportion of foreign-born nurses has grown steadily since 1998, topping 14 percent in 2003 (Brush et al., 2004).

The nursing field wants the best and the brightest undergraduate students at the top institutions to consider nursing as a possibility. It can be quite challenging to make this appeal, especially in cultures where nursing is not considered a status profession and especially at a time when many other barriers for women (particularly minority women) are falling, thereby providing new options that might be more appealing from a cultural and social standpoint.

**BOX 4-1
Summary**

Nursing's Clinical Research Workforce
- Only 0.6 percent of R.N.s are doctorally prepared.
- Between 1993 and 2002 the number of new doctorates fluctuated annually between 360 (1999) and 472 (2002).
- In 1999-2000 the mean number of years registered in doctoral programs was 8.3 for nursing graduates compared with 6.8 years for all doctoral awardees.
- The median time elapsed between entry into any graduate program to completion of the doctorate in nursing was almost twice that of other fields—15.9 years versus 8.5 years.
- The greatest nursing workforce shortage is the shortage of nursing faculty.

Increasing the Numbers of Women and Minorities
- Requires an interdisciplinary approach, because there are fields with more women and minorities that already value clinical research and an emphasis on the "lived experience" with its appreciation of participatory action (e.g., nursing, social work, psychology).
- Requires a multidisciplinary approach, because the translation of new knowledge into clinical practice and health decision making involves team or consortium building around complex problems and across institutions and sectors.

Evaluating Existing Training Efforts
- Assess the extent to which existing NIH-funded research centers are interdisciplinary and sector spanning in their training efforts (e.g., composition of advisory boards and mentors, shared courses, infrastructure supports).
- Catalog the clinical research outcomes expected of trainees at institutions with institutional research training grants, starting with the presentation of results to clinical agencies where data are collected.
- Identify the best practices of institutions that have successfully recruited and graduated minorities.

Addressing Health Disparities
- Although NINR is already focusing strongly on addressing health disparities, its success should be evaluated.

continued

BOX 4-1 Continued

- Collaborate with the National Center on Minority Health and Health Disparities (NCMHD) to develop partnership center awards to both minority-serving and research-intensive schools of nursing.

Addressing the Nursing Faculty Shortage
- Take steps to address the expected retirement over the next four to seven years of about 34 percent of nursing faculty.
- Extend the recruitment of future clinical researchers to grade school and middle school sites so that children have a vision of possibilities (e.g., Kids into Health Careers, U.S. Public Health Service) before they make choices about courses in middle school that track them.
- Discuss research, teaching, and clinical practice as different options for physicians, psychologists, nurses, pharmacists, dentists, and social workers.
- Fund continuing, seamless opportunities to nurture interest in clinical research careers (clinical and research emphases) beginning with field experiences for middle school students and continuing throughout high school and college.
- Identify and train mentors to work with young people interested in health careers and help mentors to understand clinical research career options (include school counselors).

Retaining a Representative and Diverse Workforce for Clinical Research
- Help students to find meaning in the work of a clinical researcher:
  — What does the study or research area mean to them personally? Within their cultural, ethnic group?
  — Is the problem one that matters in their community?
  — Who will benefit from their work?
- Create learning opportunities that address these questions.

Education of Future Clinical Researchers
- Expose students to interdisciplinary efforts so that they better understand complementary team members (e.g., Health Sciences Interprofessional Clinical Education Program at the University of Washington).

continued

BOX 4-1 Continued

- Have students collaborate as members of teams while learning—clinical projects, research projects.
- Broaden students' exposure beyond a single discipline (e.g., see Sung et al., 2003).
- Caution students about achieving depth in a field of study at the expense of awareness of possible connections beyond the disciplinary gaze (e.g., connecting the molecular to the organismic level).
- Alert students to the likely emergence of new disciplines (e.g., computational biology, biomedical informatics).

Resources for Educating Clinical Researchers
- T32 awards are often confined to a single discipline; they may not be broad enough to accommodate interdisciplinary clinical research.
- K30 awards—clinical research training—can support interdisciplinary training, but they are not usually structured specifically to emphasize the experiences and skill sets for research collaboration (e.g., the University of Washington Clinical Research Training Program is inclusive of physicians and other healthcare professionals).

New Paradigms in Clinical Research Training
- Interdisciplinary exposure as a foundation for team science.
- Building research networks (e.g., cross-institutional partnerships between academic health centers and less research-intensive universities, medical centers, primary care practices).
- Creative fellowship models with multisite study options for place-bound researchers, including online support.

5

Conclusions and Recommendations

Over the past two decades, policy makers and others have worried about the size and composition of the clinical research workforce, especially because of the changing demographics of the U.S. population and the concomitant implications for biomedical research and health care. Ethnic changes in the population present new challenges for understanding the health of certain populations. In addition, growing segments of the population, such as older women, will present special challenges for healthcare delivery. The increased diversity of the overall workforce, in addition to enhancing its vitality, may encourage greater participation of underrepresented minorities and women in clinical research. The benefits of their increasing representation in the clinical research workforce include greater clinical trial accrual of underrepresented minorities, robust hypothesis generation for research questions related to women and minority populations, and the likelihood that clinical research will more greatly benefit minority communities.

Unfortunately the study scope, as framed by the questions in the study charge, was much broader than the body of available data. The committee found that the first three issues in the study charge could not be answered fully because of the lack of data on the clinical research workforce. This absence of data severely limited the ability of the committee to address questions regarding supply and demand and outcome measures for existing training efforts. Data on the private sector workforce are also not available, similarly limiting the committee's ability to address the study charge about the needs of the private sector.

CONCLUSIONS AND RECOMMENDATIONS 67

The committee found, as others before them, that the single most significant impediment to achieving a better understanding of the problem was the lack of a clear, commonly agreed-upon definition of clinical research. This lack codifies the inability to set standard data definitions and will continue to hobble future attempts to understand and characterize the clinical research workforce.

Thus, the collection and analysis of data on the clinical research workforce—and clinical research overall—continues to be a challenge. More data are needed to monitor progress on the clinical research workforce; the use of standard definitions among federal agencies, careful tracking of the subsets of clinical research, and systematic evaluation of existing training efforts would be beneficial. Data standardization would also allow a better review of the composition and outcomes of study sections, which would ensure that women and minority clinical researchers are appropriately represented.

Greater numbers of physician-scientists and nurses are needed in the coming years to sustain the clinical research enterprise. Achieving these greater numbers requires examining the training and career paths for clinical research. Leaders in the field have continued to point out that the lack of a defined career path and the lengthy and costly training necessary to conduct clinical research are deterrents to entering the field. Many feel that a major and persistent obstacle to increasing the numbers of clinical researchers is the lack of regard for clinical research as a discipline in academic settings. Students who face numerous challenges to achieving career success—women and minority students face still additional challenges—may find other career paths less daunting.

More vigorous recruiting of students at earlier stages is needed to replenish the pipeline to clinical careers and in particular to reach minority populations. Various types of training programs and career tracks foster the development and retention of women and minorities in the clinical research workforce, but more programs are needed for significant improvement. Just as there are not enough data on the clinical workforce to fully understand its supply and demand, there is also not enough evaluation of existing programs to identify which ones most successfully train clinical researchers. Leaders in this field need to expand and evaluate the existing mechanisms for developing new clinical investigators, retaining investigators, and supporting mentors. They should encourage flexibility of career paths in the academic setting, as well as collaboration between basic and clinical researchers.

RECOMMENDATIONS

The study committee found that the following key themes warrant special attention in order to improve the representation of women and minorities in the clinical research workforce:

1. Adequate collection of the appropriate data;
2. Evaluation of the training landscape and mechanisms;
3. The special needs for nursing;
4. The pipeline and the career path for clinical researchers; and
5. The role of professional societies.

These themes contain systemic challenges that affect the clinical research enterprise as a whole, as well as specific challenges that should be addressed to improve the strength, character, and diversity of the workforce.

The committee offers the following recommendations for major stakeholders (federal government, academic institutions, private sector, and professional societies and associations) with the goal of strengthening and improving the diversity of the clinical research workforce.

Data Needs

A fundamental difficulty in examining issues surrounding clinical research is the lack of data on the clinical research enterprise as a whole, including data on funding levels, training programs, and who participates in the workforce. It is a challenge to examine ways to sustain and replenish the clinical research workforce when the data needed to understand the state of the clinical research enterprise are not available.

Recommendation

The National Institutes of Health of the Department of Health and Human Services should initiate a process that will develop the consistent definitions and methodologies needed to classify and report clinical research spending for all federal agencies, with advice from relevant experts and stakeholders (federal sponsors and academic centers). Such a step would allow a better understanding of the training and funding landscape and would enable accurate data collection and analysis of the clinical research workforce.

The obstacle represented by the lack of accurate data in assessing workforce needs has been noted in every edition of the congressionally mandated report to the National Institutes of Health (NIH) on the biomedical and behavioral research workforce conducted by the National Research Council. This lack stems from inconsistent definitions for classifying expenditures for clinical research across institutes and agencies (IOM, 1994). The 1997 NIH director's panel report recommended that the percentage of NIH resources devoted to clinical research, as defined by the panel, be monitored and tracked by NIH and that the results be reported annually to the NIH Director's Advisory Committee (NIH, 1997). A 2002 General Accounting Office report to Congress pointed out that NIH reports of clinical research expenditures do not have precise figures; across institutions and centers the methods that NIH uses to count clinical research dollars are inconsistent, possibly under- or overestimating its actual clinical research expenditures (GAO, 2002). Three different ways of counting clinical research dollars are used by the 20 institutions and centers that fund clinical research, producing very different results. The relevant federal agencies include NIH, Agency for Healthcare Research and Quality (AHRQ), Centers for Disease Control and Prevention (CDC), Department of Veterans Affairs (VA), Department of Defense (DOD), Health Resources and Services Administration (HRSA), and Food and Drug Administration (FDA). Federal agencies should also report on subcategories of clinical research, including preclinical or proof-of-concept human studies, Phases I-IV clinical trials, and effectiveness research (health services, outcomes, prevention, and quality research).

Training Landscape and Mechanisms: An Evaluation

Clinical research training programs are supported by public (federal government) and private (industry, foundations) sources and are implemented at academic institutions. Continued support is vital to the health of the clinical research workforce, but awareness of and access to the programs are critical if the workforce is to thrive. The effectiveness of programs should be evaluated on a regular basis.

Recommendation

The Department of Health and Human Services should work with federal clinical research sponsors to identify and describe all federally sponsored training programs (both institutional and individual) for clinical research. The information provided should identify support for each level of training and each discipline across the spectrum of clinical research (defined above). Organized links to these programs should be available on a website, including programs offered at NIH, AHRQ, VA, CDC, and HRSA. This resource should also be open to listing the institutional and individual programs offered by private sponsors for clinical research training.

The committee supports the development of the training website offered by NIH[1] and encourages NIH to modify and expand this resource to include a focus specifically on clinical research training programs.

Academic institutions should document and make publicly accessible the available programs for enhancing the participation of women and minority trainees in clinical research.

Opportunities to conduct clinical research should be well publicized at academic institutions. Information on training programs, ongoing studies, community partnerships, and fellowships should be made easily accessible and readily available to trainees.

The sponsors of federal, foundation, and industry clinical research training programs should continue to support the existing efforts to train, develop, and sustain the careers of clinical researchers.

The pharmaceutical industry is the largest sponsor of clinical trials, and the recruitment of participants, particularly in underrepresented communities, has been a consistent concern. Racial and ethnic diversity of the workforce is an important factor in improving participation in clinical research in these communities. Industry-sponsored fellowships for master's programs in clinical trial management for undergraduate nurses and students of underrepresented minority students would be one way to better engage minority communities in the clinical research enterprise.

---

[1] NIH at http://www.training.nih.gov/careers/careercenter. Accessed on November 22, 2004.

Recommendation

Federal sponsors (NIH, CDC, AHRQ, VA, DOD) should ensure adequate representation of women and minorities in study section review panels that review clinical research.

A system that ensures adequate representation of clinical researchers, women, and minorities will help federal research sponsors to maintain an appropriate balance in the review process. This finding was reinforced in the 1997 NIH director's panel report, which recommended that NIH ensure fair and effective reviews of extramural grant applications. Panels that review clinical research should have a significant proportion of experienced clinical investigators (NIH, 1997).

Recommendation

Federal agencies and academic institutions should periodically evaluate how well their current training programs are enhancing the racial and ethnic diversity of trainees and they should modify these programs as needed to increase the programs' effectiveness in clinical research.

As the demographics of the student population change, so may training needs. Academic institutions need to assess their training programs on a regular basis so that changes can be made if necessary to ensure success in pursuing clinical research careers.

Nursing Professionals

The continuing shortfall of nursing professionals is compounded in clinical research by the longer time required for specialized training, and the fewer numbers of nursing faculty involved in clinical research.

Recommendation

The need for appropriately trained nursing professionals in the clinical research workforce is especially urgent. A significant push is needed to increase the numbers of minorities entering the nursing profession. Additional attention should be paid to the clinical research training of nurse-scientists, nursing students, and nursing faculty at all academic levels.

The shortage of practicing nurses in the U.S. healthcare delivery system presents challenges to training nurse-scientists. Nursing faculty are pressed to be fully engaged in training nursing students, leaving less time for clinical research. New approaches should be explored to prepare both medical and graduate nursing students in a core interdisciplinary research curriculum to foster interdisciplinary collaboration (e.g., social work, epidemiology). Training initiatives already underway by the National Institute for Nursing Research and the American Academy of Nursing to strengthen nursing school curricula across all levels should be supported and enhanced. Several options for enhancement of training efforts follow.

- Expansion of fast-track B.S.N.-to-doctoral study to reduce the time to career launch;
- Expansion of existing alternative career programs (e.g., entry to practice for non-nurses, accelerated B.S.N.-to-master's programs) for diploma nurses;
- Summer programs in clinical research for undergraduate and master's nursing students;
- Expanded support for HRSA Division of Nursing training grants for specialized doctoral programs in the schools of nursing affiliates of academic medical centers;
- Institutional pre- and postdoctoral National Research Service Awards in clinical research for nurses that require and support funding for mentorship by a clinical research investigator;
- Foundation- and industry-supported scholars programs like the John A. Hartford Foundation's program for Building Academic Geriatric Nursing Capacity and the Pfizer Postdoctoral Fellowships in Nursing Research;
- K12 fellowship set-asides in clinical research for academic nursing faculty; and
- Sabbaticals for midcareer nurse-scientists in clinical research.

Replenishing the Pipeline: A Flexible Career Path

Given the long training period required for clinical research, entry points throughout a clinical research career path, not just at trainee levels, could increase the workforce. Additional efforts are needed to retain scientists in the clinical research workforce.

Recommendation

Federal sponsors of clinical research should amplify the existing funding mechanisms and create new ones that allow flexibility in career training, such as second-career programs, reentry mechanisms, and service payback agreements. These programs should be described on the NIH training website. In addition, other entry routes into the clinical research path, including short-term training programs, should be developed.

Given the length of time required for training, the clinical research career pathway is fixed, with little latitude for alternative entry points. This situation may filter out well-qualified candidates from other biomedical research career tracks who could enter clinical research in shorter training programs (e.g., one-year programs for medical students). Career paths for women and minorities may not follow the conventional, rigid model because of considerations such as family responsibilities, debt- and risk-aversive trends, and differential debt burdens among different communities. Flexibility in career training is essential if diverse candidates are to thrive in clinical research careers.

Recommendation

Academic institutions should develop strategies to attract mentors and reward mentorship in clinical research training. A special emphasis should be placed on the women and minorities who carry the greatest burden of mentorship responsibilities for women and minority scientists.

Academic institutions should develop flexibility so that the time required for tenure reflects the time course for research, particularly for physician-scientists. The process should also recognize individual differences, academic contribution, and academic service, including mentoring, and where possible should use quantifiable measures of excellence. Academic service (e.g., institutional review boards and other committee service) should count as a significant positive factor in decisions relating to promotion and tenure. Efforts should be made to improve the composition of promotion and tenure committees so that women and minority clinical researchers are represented.

Individuals in clinical research pathways should be given the necessary infrastructure to achieve success, including clearly defined benchmarks.

Clear distinctions should be made between the clinical service role and that of investigators with independent research resources. Credit should be given for volunteer efforts that foster science and math education in the K-12 environment.

The Role of Professional Societies

Professional societies play a major role in the scientific community, as publishers of journals, sponsors of awards, and representatives of their scientific community.

Recommendation

Specialty medical and nursing societies should form a new consortium that would assume an enhanced role in fostering a diverse clinical research workforce.

The consortium could be based on the Federation of American Societies for Experimental Biology or American Heart Association models and should focus on common development, implementation, and advocacy regarding clinical research training and clinical research priorities (Burroughs Wellcome Fund, 2003). These efforts should include an emphasis on the training, retention, and advancement of women and minorities:

- Societies should give high priority to developing clinical researchers.
- Societies should encourage, promote, and foster mentoring of clinical research trainees, paying particular attention to women and minorities.
- Societies should develop resource sharing and facilitate interaction to foster clinical research training programs, mentors, and trainees. Societies should work toward intersociety initiatives.
- Clinical research training mechanisms should include identification and creation of a database at the pipeline level whereby potential trainee candidates can be identified and mentors assigned at the earliest possible level.
- Societies should develop or institute tracking mechanisms to determine training effort outcomes and retention of trainees and mentors.

References

AACN (American Association of Colleges of Nursing). 2002a. Faculty resignations and retirements (unpublished data). Washington, D.C.: AACN.

AACN. 2002b. Institutional data systems, 1990-2002. Washington, D.C.: AACN.

AACN. 2003. Faculty shortages in baccalaureate and graduate nursing programs: Scope of the problem and strategies for expanding the supply. Online at http://www.aacn.nche.edu/Publications/WhitePapers/FacultyShortages.htm. Accessed February 9, 2004.

AACN and the National Organization of Nurse Practitioner Faculties. 2000. 1999-2000 enrollment in baccalaureate and graduate programs in nursing. Online at http://www.aacn.nche.edu/Media/pdf/SullivanReport.pdf. Accessed September 2004.

AAMC (Association of American Medical Colleges). 1999. For the Health of the Public: Ensuring the Future of Clinical Research. Washington, D.C.: AAMC.

AAMC. 2002. AAMC faculty roster system. Online at http://www.aamc.org/data/facultyroster/reports.htm. Accessed January 9, 2004.

AAMC. 2003. Medical school graduation questionnaire. Washington, D.C.: AAMC.

AAMC. 2004a. Medical school tuition and young physician indebtedness. Online at https://services.aamc.org/Publications/showfile.cfm?file=version21.pdfandCFID=123849and CFTOKEN=4e4244a-164c1fe3-1be7-4e02-bbaa-306212f0085f. Accessed April 19, 2004.

AAMC. 2004b. Women in U.S. academic medicine: Statistics and medical school benchmarking, 2003-2004. Washington, D.C.: AAMC.

ACP (American College of Physicians). 1991. Promotion and tenure of women and minorities on medical school faculties. Annals of Internal Medicine 114:63-68. Online at http://www.aacn.nche.edu and http://www.nonpf.com. Accessed February 9, 2004.

Adler, R. 2001. Women and profits. Harvard Business Review 79:30.

Ahrens, E. H. 1992. The Crisis in Clinical Research. New York: Oxford University Press.

Airhihenbuwa, C. O., S. Kumanyika, T. D. Agurs, A. Lowe, D. Saunders, and C. B. Morssink. 1996. Cultural aspects of African American eating patterns. Ethnicity and Health 1:245-260.

Allen, R. S., and K. A. Montgomery. 2001. Applying an organizational development approach to creating diversity. Organizational Dynamics 30:149-161.

American Cancer Society, Burroughs Wellcome Fund, Howard Hughes Medical Institute. 2002. The role of the private sector in training the next generation of biomedical scientists. Online at http://www.bwfund.org/news/special_reports/Funders percent20Report.pdf. Accessed September 17, 2003.

Anderson, R. M. 1995. Patient empowerment and the traditional medical model: A case of irreconcilable differences? Diabetes Care 18:412-415.

Barzansky, B., and S. I. Etzel. 2001. Educational programs in U.S. medical schools, 2000-2001. Journal of the American Medical Association 286:1049-1055.

Barzansky, B., and S. I. Etzel. 2002. Educational programs in U.S. medical schools, 2001-2002. Journal of the American Medical Association 288:1067-1072.

Barzansky, B., and S. I. Etzel. 2003. Educational programs in U.S. medical schools, 2002-2003. Journal of the American Medical Association 290:1190-1196.

Beasley, B. W., S. M. Wright, J. Cofrancesco, S. F. Babbott, P. A. Thomas, and E. B. Bass. 1997. Promotion criteria for clinician-educators in the United States and Canada. Journal of the American Medical Association 278:723-728.

Berger, J. T. 1998. Culture and ethnicity in clinical care. Archives of Internal Medicine 158:2085-2090.

Berlin, L. E., G. D. Bednash, and O. Alsheimer. 1993. 1992-1993 Enrollment and Graduations in Baccalaureate and Graduate Programs in Nursing. Washington, D.C.: American Association of Colleges of Nursing.

Berlin, L. E., J. Stennett, and G. D. Bednash. 2003. 2002-2003 Enrollment and Graduations in Baccalaureate and Graduate Programs in Nursing. Washington, D.C.: American Association of Colleges of Nursing.

Bickel, J. 2001. Gender equity in undergraduate medical education: A status report. Journal of Women's Health and Gender-Based Medicine 10:261-270.

Bickel, J. 2004. Advancing women in academic medicine. In: Achieving XXcellence in Science, ed., S. Shaywitz and J. Hahm, pp. 38-42. Washington, D.C.: The National Academies Press.

Bickel, J., D. Wara, B. F. Atkinson, et al. Association of American Medical Colleges Project Implementation Committee. 2002. Increasing women's leadership in academic medicine: Report of the AAMC Project Implementation Committee. Academic Medicine 77:1043-1061.

Bloom, F. E. 2003. Presidential address. Science as a way of life: Perplexities of a physician-scientist. Science 300:1680-1685.

Bonow R. O., and S. C. Smith. 2004. Cardiovascular manpower: The looming crisis. Circulation 109:817-820.

Bowen, W. G., and D. Bok. 1998. The Shape of the River: Long Term Consequences of Considering Race in College and University Admissions. Princeton, N.J.: Princeton University Press.

Brush, B. L., J. Sochalski, and A. M. Berger. 2004. Imported care: Recruiting foreign nurses to U.S. health care facilities. Health Affairs 23:78-87.

REFERENCES

Buckley, L. M., K. Sanders, M. Shih, and C. Hampton. 2000a. Attitudes of clinical faculty about career progress, career success and recognition, and commitment to academic medicine. Archives of Internal Medicine 160:2625-2629.

Buckley, L. M., K. Sanders, M. Shih, S. Kallar, and C. Hampton. 2000b. Obstacles to promotion? Values of women faculty about career success and recognition. Academic Medicine 75:283-288.

Buerhaus, P., D. Staiger, and D. Auerbach. 2000. Implications of an aging registered nurse workforce. Journal of the American Medical Association 283:2948-2954.

Bureau of Labor Statistics, U.S. Department of Labor. 1999. Millenial Themes: Age, Education, Services. Originally published December 1, 1999, as MLR: The Editor's Desk. Online at http//:www.bls.gov/opub/ted/1999/nov/wk5/art03.htm. Accessed October 4, 2003.

Burroughs Wellcome Fund. 2003. Challenges facing the clinical research enterprise: The response of medical specialty and clinical research societies. September 2003. Online at http://www.bwfund.org/programs/translational/special_reports.html. Accessed June 20, 2005.

Cabana, M. D., C. S. Rand, N. R. Powe, et al. 1999. Why don't physicians follow clinical practice guidelines? Journal of the American Medical Association 282:1458-1465.

Campbell, E. G., J. S. Weissman, E. Moy, and D. Blumenthal. 2001. Status of clinical research in academic health centers: Views from the research leadership. Journal of the American Medical Association 286:800-806.

Cantor, J. C., E. L. Miles, L. C. Baker, and D. C. Barker. 1996. Physician service to the underserved: Implications for affirmative action in medical education. Inquiry 33:167-181.

Carr, P. L., A. S. Ash, R. H. Friedman, et al. 1998. Relation of family responsibilities and gender to the productivity and career satisfaction of medical faculty. Annals of Internal Medicine 129:579-580.

CenterWatch. 2001. An Industry in Evolution. Boston, Mass.: CenterWatch.

Chan, J. M., K. S. Roth, and I. B. Salusky. 2002. Clinical research and training: An overview. Journal of Pediatrics 140:293-298.

Cheung, V. G., N. Nowak, W. Jang, I. R. Kirsch, et al. Resource Consortium. 2001. Integration of cytogenetic landmarks into the draft sequence of the human genome. Nature 409:953-958.

Christman, L. 1998. Who is a nurse? J Nurs Sch 30:211-214.

Church, E. 2001. Market favours firms with women at top. Online at http://www.healthsmith.com/Org%20well/womenontop.htm. Accessed August 11, 2004.

Clark, J. 1999. Minorities in Science and Math. Columbus, Ohio: ERIC Clearinghouse for Science, Mathematics, and Environmental Education.

Collins, F. S., M. Morgan, and A. Patrinos. 2003. The Human Genome Project: Lessons from large-scale biology. Science 300:286-290.

Corbie-Smith, G., S. B. Thomas, M. V. Williams, and S. Moody-Ayers. 1999. Attitudes and beliefs of African Americans toward participation in medical research. Journal of General Internal Medicine 14:537-546.

Cox, T. H. 1993. Cultural Diversity in Organizations: Theory, Research and Practice. San Francisco: Berrett-Koehler.

Crowley, S., D. Fuller, W. Law W, et al. 2004. Improving the climate in research and scientific training environments for members of underrepresented minorities. Neuroscientist 10:26-30.

DePaolo, L. V., and P. C. Leppert. 2002. Providing research and research training infrastructure for clinical research in the reproductive science. American Journal of Obstetrics and Gynecology 187:1087-1090.

Division of Nursing, Bureau of Health Professions, Health Resources and Services Administration. 2001. The registered nurse population: National sample survey of registered nurses. Unpublished special reports generated for the American Association of Colleges of Nursing.

Drenth, J. P. 1998. Multiple authorship: The contribution of senior authors. Journal of the American Medical Association 280:219.

Elliott, E. D. 2001. The genome and the law: Should increased genetic knowledge change the law? Harvard Journal of Law and Public Policy 25:61-70.

Fang, D., and R. E. Meyer. 2003a. Effect of two Howard Hughes Medical Institute research training programs for medical students on the likelihood of pursuing research careers. Academic Medicine 78:1271-1280.

Fang, D., and R. E. Meyer. 2003b. Ph.D. faculty in clinical departments of U.S. medical schools, 1981-1999: Their widening presence and roles in research. Academic Medicine 78:167-176.

Ferraro, G., and L. Martin. 2000. Reaping the bottom line benefits of diversity. Online at http://www.gwsae.org/ExecutiveUpdate/2000/July/BottomLineBenefits.htm. Accessed August 11, 2004.

Fey, M. F. 2002. Impact of the Human Genome Project on the clinical management of sporadic cancers. Lancet Oncology 3:349-356.

Fox, M. F. 1995. Publication, performance and reward in science and scholarship. In: Higher Education: Handbook of Theory and Research, vol. 1, ed. J. C. Smart. New York: Agathon Press.

GAO (General Accounting Office). 2002. The GAO report to congressional committees on clinical research: NIH has implemented key provisions of the Clinical Research Enhancement Act. September 20. Online at http://www.gao.gov/new.items/d02965.pdf. Accessed June 20, 2005.

Gartland, J. J., M. Hojat, E. B. Christian, C. A. Callahan, and T. J. Nasca. 2003. African American and white physicians: A comparison of satisfaction with medical education, professional careers, and research activities. Teaching and Learning in Medicine 15:106-112.

Gelijns, A. C., and S. O. Thier. 2002. Medical innovation and institutional interdependence: Rethinking university-industry connections. Journal of the American Medical Association 287:72-77.

Gerling, I. C., S. S. Solomon, and M. Bryer-Ash. 2003. Genomes, transcriptomes, and proteomes: Molecular medicine and its impact on medical practice. Archives of Internal Medicine 163:190-198.

Gifford, A. L., W. E. Cunningham, K. C. Heslin, R. M. Andersen, T. Nakazono, D. K. Lieu, M. F. Shapiro, and S. A. Bozzette. 2002. HIV Cost and Services Utilization Study Consortium. 2002. Participation in research and access to experimental treatments by HIV-infected patients. New England Journal of Medicine 346:1373-1382.

Giuliano, A. R., N. Mokuau, C. Hughes, et al. 2000. Participation of minorities in cancer research: The influence of structural, cultural, and linguistic factors. Annals of Epidemiology 10(8 suppl):S22-S34.

Goldman, E., and E. Marshall. 2002. NIH grantees: Where have all the young ones gone? Science 298:40-41.

Goldstein, J. L., and M. S. Brown. 1997. The clinical investigator: Bewitched, bothered, and bewildered—but still beloved. Journal of Clinical Investigations 99:2803-2812.

Goldstein, J. L., and M. S. Brown. 2002. Acceptance remarks of Joseph L. Goldstein and Michael S. Brown on the occasion of the presentation of the Kober Medal, 2002. Journal of Clinical Investigations 110:S11-S13.

Gordon, S. M., M. W. Heft, R. A. Dionne, et al. 2003. Capacity for training in clinical research: Status and opportunities. Journal of Dental Education 67:622-629.

Guelich, J. M., B. H. Singer, M. C. Castro, and L. E. Rosenberg. 2002. A gender gap in the next generation of physician-scientists: Medical student interest and participation in research. Journal of Investigative Medicine 50:412-418.

Gurin, P. 1999. Expert report of Patricia Gurin, Gratz et al. v. Bollinger et al., No. 97-75321 (E.D. Mich.); Grutter et al. v. Bollinger et al., No. 97-75928 (E.D. Mich.). In: The Compelling Need for Diversity in Higher Education, pp. 99-234. Ann Arbor: University of Michigan, Office of the Vice President and General Counsel.

Haynes, M. A. 1999. Testimony of Alfred Haynes, MD, Before the Labor, Health and Human Services, and Education Subcommittee, Senate Committee on Appropriations, U.S. Senate, 106th Cong., 1st sess., January 21. Online at http://www7.nationalacademies.org/ocga/testimony/Cancer_Research_Among_Minorities.asp#TopOfPage. Accessed June 10, 2005.

Heinig, S. J., A. S. Quon, R. E. Meyer, and D. Korn. 1999. The changing landscape for clinical research. Academic Medicine 74:726-745.

Hetteberg, C., and C. A. Prows. 2004. A checklist to assist in the integration of genetics into nursing curricula. Nursing Outlook 52:85-88.

HHMI (Howard Hughes Medical Institute). 2003a. Research training fellowships for medical students. Online at http://www.hhmi.org/grants/funding/comp_annc/2004_med_pa.pdf. Accessed November 25, 2003.

HHMI. 2003b. The HHMI-NIH Research Scholars Program. Online at http://www.hhmi.org/research/cloister. Accessed November 25, 2003.

Horner, S.D., E. Abel, K. Taylor, and D. Sands. 2004. Using theory to guide the diffusion of genetics content in nursing curricula. Nursing Outlook 52:80-84.

HRSA (Health Resources and Services Administration). 2000. The pharmacist workforce: A study of the supply and demand for pharmacists. Online at ftp://ftp.hrsa.gov/bhpr/nationalcenter/pharmacy/pharmstudy.pdf. Accessed January 14, 2004.

HRSA, Bureau of Health Professions. 2002. Projected supply, demand, and shortages of registered nurses: 2000-2020. Online at http://bhpr.hrsa.gov/healthworkforce/reports/rnproject/report.htm. Accessed August 11, 2004.

HRSA, Bureau of Health Professions. 2003. The registered nurse population 1980-2000 survey. Online at http://bhpr.hrsa.gov/healthworkforce/reports/rnsurvey/rnss1.htm. Accessed August 11, 2004.

Hunt, L. M., J. Pugh, and M. Valenzuela. 1998. How patients adapt diabetes self-care recommendations in everyday life. Journal of Family Practice 46:207-215.

IOM (Institute of Medicine). 1994. Careers in Clinical Research: Obstacles and Opportunities. Washington, D.C.: National Academy Press.
IOM. 2001a. Exploring Biological Contributions to Human Health: Does Sex Matter? Washington, D.C.: National Academy Press.
IOM. 2001b. The Right Thing to Do, the Smart Thing to Do: Enhancing Diversity in Health Professions. Washington, D.C.: National Academy Press.
IOM. 2001c. Small Clinical Trials: Issues and Challenges. The Future of Public Health. Washington, D.C.: National Academy Press.
IOM. 2004a. In the Nation's Compelling Interest: Ensuring Diversity in the Health Care Workforce. Washington, D.C.: The National Academies Press.
IOM. 2004b. Keeping Patients Safe: Transforming the Work Environment of Nurses. Washington, D.C.: The National Academies Press.
Juliano, R. L., and G. S. Oxford. 2001. Critical issues in Ph.D. training for biomedical scientists. Academic Medicine 76:1005-1012.
Kaplan, S. H., L. M. Sullivan, K. A. Dukes, C. F. Phillips, R. P. Kelch, and J. G. Schaller. 1996. Sex differences in academic advancement: Results of a national study of pediatricians. New England Journal of Medicine 335:1282-1289.
Kempers, R. D. 2001. Ethical issues in biomedical publications. Hum Fertil (Camb) 4:261-266.
Kenkre, J. E., and D. R. Foxcroft. 2001. Career pathways in research: Pharmaceutical. Nursing Standard 16:36-39
Killien, M., J. A. Bigby, V. Champion, et al. 2000. Involving minority and underrepresented women in clinical trials: The National Centers of Excellence in Women's Health. Journal of Women's Health and Gender Based Medicine 9:1061-1070.
Komaromy, M., K. Grumbach, M. Drake, et al. 1996. The role of black and Hispanic physicians in providing health care for underserved populations. New England Journal of Medicine 334:1305-1310.
Kotchen, T. A., T. Lindquist, K. Malik, and E. Ehrenfeld. 2004. NIH peer review of grant applications for clinical research. Journal of the American Medical Association 291(7):836-843.
Kowalsky, S. F. 1996. Opportunities for pharmacists in clinical research. Pharm Pract Manage Q 16:1-3.
Krakower, J. Y., D. J. Williams, and R. F. Jones. 1999. Review of U.S. medical school finances, 1997-1998. Journal of the American Medical Association 282:847-854.
Lee, P. Y., K. P. Alexander, B. G. Hammill, S. K. Pasquali, and E. D. Peterson. 2001. Representation of elderly persons and women in published randomized trials of acute coronary syndromes. Journal of the American Medical Association 286:708-713.
Ley, T. J., and L. E. Rosenberg. 2002. Removing career obstacles for young physician-scientists—Loan-repayment programs. New England Journal of Medicine 346:368-372.
Lippman, H. 2000. Variety is the spice of a great workforce. Business and Health 18:24-29.
Mark, S., and J. Gupta. 2002. Reentry into clinical practice: Challenges and strategies. Journal of the American Medical Association 288:1091-1096.
Martinez, R. 2003. Workshop Presentation: Opportuntities to Address Clincial Research Workforce Diversity Needs.

REFERENCES

Mateo, M. A., and S. P. Smith. 2003. Workforce diversity in hospitals. Nurs Leadership Forum 7:143-149.

McBride, A. B. 1999. Breakthroughs in nursing education: Looking back, looking forward. Nursing Outlook 47:114-119.

McBride, A. B., 2003. Workshop Presentation: Opportuntities to Address Clincial Research Workforce Diversity Needs.

McCracken, D. 2000. Winning the talent war for women: Sometimes it takes a revolution. Harvard Business Revieew 78:159-166.

McGivern, D. O. 2003. The scholars' nursery. Nursing Outlook 51:59-64.

McGuire, L. K., M. R. Bergen, and M. L. Polan. 2004. Career advancement for women faculty in a U.S. school of medicine: Perceived needs. Academic Medicine 79:319-325

Mervis, J. 2003. NIH program gives minorities a chance to make their MARC. Science 301:455-456.

Mike, V. 2003. Evidence and the future of medicine. Eval Health Prof 26:127-152.

Miller, E. D. 2001. Clinical investigators—the endangered species revisited. Journal of the American Medical Association 286:845-846.

Morahan, P. S., M. L. Voytko, S. Abbuhl, et al. 2001. Ensuring the success of women faculty at AMCs: Lessons learned from the National Centers of Excellence in Women's Health. Academic Medicine 76:19-31.

Morahan, P., and J. Bickel. 2002. Capitalizing on women's intellectual capital. Academic Medicine 77:110-112

Moskowitz, J., and J. N. Thompson. 2001. Enhancing the clinical research pipeline: Training approaches for a new century. Academic Medicine 76:307-315.

Mueller, K. 1998. Diversity and the bottom line. Harvard Business Review.

Nathan, D. G., and J. D. Wilson. 2003. Clinical research and the NIH: A report card. New England Journal of Medicine 349:1860-1865.

National League for Nursing. 2003. Nurse Educators 2002: Report of the Faculty Census Survey of R.N. and Graduate Programs. New York: National League for Nursing.

National Opinion Research Center. 2001. Survey of earned doctorates. Unpublished special reports generated for the American Association of Colleges of Nursing. Chicago: National Opinion Research Center.

NSF (National Science Foundation). 2004. Women, minorities, and persons with disabilities in science and engineering. Online at http://www.nsf.gov/sbe/srs/wmpd/start.htm. Accessed August 9, 2004.

Newton, D. A., and M. S. Grayson. 2003. Trends in career choices by U.S. medical school graduates. Journal of the American Medical Association 290:1179-1182.

NIH (National Institutes of Health). 1992. Women in biomedical careers: Dynamics of change, strategies for the 21st century. Office of Research on Women's Health.

NIH. 1994. NIH guidelines on the inclusion of women and minorities as subjects in clinical research. Federal Register 59:14508-14513.

NIH. 1997. Director's Panel on Clinical Research: Report to the Advisory Committee to the NIH Director. Online at http://www.nih.gov/news/crp/2report.htm. Accessed August 11, 2004.

NIH. 1998. Number of applications to NIH for competing research project grants, by degree of investigator, FY 1986-95. Online at http://www.cmwf.org/programs/taskforc/cwf_ahc_bench_fig2_312.asp#figure12. Accessed September 12, 2003.

NIH. 1999. AXXS: Achieving XXcellence in Science. Office of Research on Women's Health.
NIH. 2002a. NIH Clinical Research: Competing Awards. Online at http://grants1.nih.gov/grants/award/trends/clincomp9601.htm.
NIH. 2002b. Underrepresented minority physician-scientists: Today's obstacles; tomorrow's opportunities—executive summary. Online at: http://minorityopportunities.nci.nih.gov/resources/md-exec-summ.pdf.
Nonnemaker, L. 2000. Women physicians in academic medicine: New insights from cohort studies. New England Journal of Medicine 342:399-405.
NRC (National Research Council). 2000. Addressing the Nation's Changing Needs for Biomedical and Behavioral Scientists. Washington, D.C.: National Academy Press.
NRC. 2001. From Scarcity to Visibility: Gender Differences in the Careers of Doctoral Scientists and Engineers, ed. S. J. Long. Panel for the Study of Gender Differences in Career Outcomes of Science and Engineering Ph.D.'s. Washington, D.C.: National Academy Press.
NRC. 2004. Achieving XXcellence in Science. Role of Professional Societies in Advancing Women in Science. Washington, D.C.: The National Academies Press.
NRC. 2005a. Advancing the Nation's Health Needs: NIH Research Training Programs. Washington, D.C.: The National Academies Press.
NRC. 2005b. Assessment of NIH Minority Research and Training Programs: Phase 3. Washington, D.C.: The National Academies Press.
NRMP (National Resident Matching Program). 2003. National resident matching program results and data. Online at http://www.nrmp.org/res_match/data_tables.html. Accessed August 9, 2004.
Nugent, K. E., G. Childs, R. Jones, and P. Cook. 2004. A mentorship model for the retention of minority students. Nursing Outlook 52:89-94.
Pendharkar, S. R. 2003. Medical women in academia: Silenced by the system. CMAJ 168:542-544; author reply 544.
PhRMA. 2004. 2004 industry profile. Online at http://www.phrma.org/publications/publications//2004-03-31.937.pdf. Accessed June 3, 2004.
Phillips, D. F. 2000. IRBs search for answers and support during a time of institutional change. Journal of the American Medical Association 283:729-730.
Randal, J. 2001. Examining IRBs: Are review boards fulfilling their duties? Journal of the National Cancer Institute 93:1440-1441.
Reame, N. 2003. Workshop Presentation: Opportunities to Address Clinical Research Workforce Diversity Needs.
Saha, S., M. Komaromy, T. D. Koepsell, and A. B. Bindman. 1999. Patient-physician racial concordance and the perceived quality and use of health care. Archives of Internal Medicine 159:997-1004.
Sessa, V., and J. Taylor. 2000. Executive Selection: Strategies for Success. Greensboro, N.C.: Center for Creative Leadership.
Snyderman, R. 2004. The clinical researcher: An "emerging" species. Journal of the American Medical Association 291:882-883.
Socolar, R. S., L. S. Kelman, C. M. Lannon, and J. A. Lohr. 2000. Institutional policies of U.S. medical schools regarding tenure, promotion, and benefits for part-time faculty. Academic Medicine 75:846-849.

Solomon, S. S., S. C. Tom, J. Pichert, D. Wasserman, and A. C. Powers. 2003. Impact of medical student research in the development of physician-scientists. Journal of Investigative Medicine 51:149-156.

Spratley, E., A. Johnson, J. Sochalski, M. Fritz, and W. Spencer. 2000. The registered nurse population: Findings from the National Sample Survey of Registered Nurses. Department of Health and Human Services, Health Resource and Services Administration. Online at http://bhpr.hrsa.gov/healthworkforce/reports/rnsurvey/rnss1.htm. Accessed November 24, 2003.

Stashenko, P., R. Niederman, and D. DePaola. 2002. Basic and clinical research: Issues of cost, manpower needs, and infrastructure. Journal of Dental Education 66:927-938 (discussion 939-941).

Sung, N. S., W. F. Crowley, M. Genel, et al. 2003. Central challenges facing the national clinical research enterprise. Journal of the American Medical Association 289:1278-1287.

Thomas, P. A., M. Diener-West, M. I. Canto, D. R. Martin, W. S. Post, and M. B. Streiff. 2004. Results of an academic promotion and career path survey of faculty at the Johns Hopkins University School of Medicine. Academic Medicine 79:258-264.

Thomson CenterWatch. 2004. Shifts in the foundation of drug development. Thomson CenterWatch 11:1.

VA (Department of Veterans Affairs). 2003. Health Services Research and Development Service. Online at http://www.hsrd.research.va.gov. Accessed January 9, 2004.

Williams, J. K., T. Trip-Reimer, D. Schutte, and J. J. Barnette. 2004. Advancing genetic nursing knowledge. Nursing Outlook 52:73-79.

Wolf, M. 2002. Clinical research career development: The individual perspective. Academic Medicine 77:1084-1088.

Woods, N. 2003. Workshop Presentation: Opportunties to Address Clincial Research Workforce Diversity Needs.

Wright A. L., L. A. Schwindt, T. L. Bassford, et al. 2003. Gender differences in academic advancement: Patterns, causes, and potential solutions in one U.S. college of medicine. Academic Medicine 78:500-508.

Wyngaarden, J. B. 1979. The clinical investigator as an endangered species. New England Journal of Medicine 301:1254-1259.

Yates, J. 2003. Workshop Presentation: Opportunities to Address Clinical Research Workforce Diversity Needs.

Yedidia, M. J., and J. Bickel. 2001. Why aren't there more women leaders in academic medicine? The views of clinical department chairs. Academic Medicine 76:453-465.

Zemlo, T. R., H. H. Garrison, N. C. Partridge, and T. J. Ley. 2000. The physician-scientist: Career issues and challenges at the year 2000. FASEB Journal 14:221-230.

Zerhouni, E. 2003. The NIH roadmap. Science 302:63-72.

Appendixes

Appendix A

Biographies of Speakers

CLAUDIA R. BAQUET, M.D., M.P.H., is associate dean for policy and planning and associate professor of epidemiology and preventive medicine at the University of Maryland School of Medicine. She also serves as director of the Maryland Area Health Education Center, director of the Center for Health Policy/Health Services Research, director of the Cancer Disparities and Intervention Research Program, principal investigator of the Maryland Special Populations Cancer Research Network, and director of the University of Maryland Statewide Health Network. Throughout her government and academic career, Dr. Baquet has been a champion of issues related to health disparities and the underserved and is considered a leading national expert on cancer in minority and low-income populations.

LOIS COLBURN is assistant vice president in the Division of Community and Minority Programs of the Association of American Medical Colleges (AAMC). She is currently deputy director of the Robert Wood Johnson Foundation–Kellogg Health Professions Partnership Initiative, which helps to develop partnerships among academic medical centers, undergraduate institutions, and secondary schools as a means of increasing the number of academically competitive minority students in the health professions pipeline. She is also the editor of *Minorities in Medical Education: Facts and Figures,* an annual publication detailing the enrollment and graduation trends of minority students in U.S. medical schools. Ms. Colburn is involved as well in the development of minority faculty initiatives, most notably the AAMC Health Services Research Institute.

WILLIAM CROWLEY JR., M.D., is director of clinical research and chief of the Reproductive Endocrine Unit at Massachusetts General Hospital, director of the National Center for Infertility Research, and professor of medicine at Harvard University. Dr. Crowley is the founder of the Academic Health Center Clinical Research Forum and is a member of the board of directors and executive committee of the Federation of American Societies for Experimental Biology. His research interests are neuroendocrine control and reproduction and growth, physiology of puberty, and physiology of gonadotropin secretion.

SHERINE E. GABRIEL, M.D., M.Sc., is professor of epidemiology and medicine at Mayo Medical School and is currently chair of the Department of Health Sciences Research at Mayo Clinic, Rochester, Minnesota. She holds dual appointments as Mayo Clinic consultant in the Departments of Internal Medicine/Rheumatology and Health Sciences Research/Epidemiology. Her research has been widely recognized, nationally and internationally. Dr. Gabriel's commitment to clinical research also extends to clinical research training. In 1999 she led a team of clinical investigators from Mayo who prepared and submitted a proposal for a new clinical research training program at Mayo Clinic in response to the new K30 initiative. This grant was awarded and the Mayo application received a score of 133, establishing Mayo Clinic as one of the top medical centers receiving this institutional award.

WILLIAM R. GALEY, Ph.D., is director of the Graduate Science Education Program at the Howard Hughes Medical Institute (HHMI). Before joining HHMI, Dr. Galey was at the University of New Mexico School of Medicine, where he served as interim associate dean for research, assistant dean for graduate studies, and director of the Biomedical Sciences Graduate Program, a training program for Ph.D. and M.D.-Ph.D. candidates.

E. NIGEL HARRIS, M.Phil., M.D., D.M., is dean and senior vice president for academic affairs of the Morehouse School of Medicine, a position he has held since January 1996. Dr. Harris's research career has been largely devoted to the study of antiphospholipid antibodies. He helped devise the anticardiolipin test, and later introduced calibrators for the anticardiolipin assay and units for measurement of anticardiolipin antibody levels. In 1987 Dr. Harris established a new laboratory with Dr. Silvia Pierangeli.

WILLIAM N. KELLEY, M.D., former chief executive officer and dean of the University of Pennsylvania Health System and School of Medicine, is professor of medicine and of biochemistry and biophysics at the University of Pennsylvania. He also currently serves as a director of Merck & Co., Beckman Coulter Inc., GenVec Inc., and Advanced Bio-Surfaces Inc., and as a trustee of Emory University and the Woodruff Health Sciences Center of Emory University. Dr. Kelley is a member of the Institute of Medicine.

EVAN D. KHARASCH, M.D., Ph.D., is the assistant dean for clinical research at the University of Washington School of Medicine. He is also professor and director of research, Department of Anesthesiology, and adjunct professor of medicinal chemistry at the University of Washington, Seattle. His research areas include clinical pharmacology of anesthetic and analgesic drugs; laboratory, clinical, and noninvasive assessment of drug disposition, metabolism, and drug interactions; clinical optimization of analgesic drug use; mechanisms of interindividual variability in opioid disposition and response; and mechanisms of anesthetic toxification and detoxification. Dr. Kharasch is a practicing anesthesiologist.

THOMAS J. LAWLEY, M.D., is dean and William P. Timmie Professor of Dermatology at Emory University School of Medicine. He is an internationally known expert in autoimmune skin diseases. Dr. Lawley currently serves on the Administrative Council of the Association of American Medical Colleges. He is president of the Emory Medical Care Foundation (Emory's physician practice plan at Grady Hospital) and president of the Emory Children's Center.

MARY D. LEVECK, R.N., Ph.D., is the deputy director of the National Institute of Nursing Research (NINR) at the National Institutes of Health (NIH). She also serves as the director of the Division of Extramural Activities. Prior to assuming her current duties, she was a branch chief and extramural program director at NINR beginning in 1990. Previously, she held faculty and administrative positions at the College of Nursing, University of South Carolina, Columbia. At NIH her major initiatives have been in the area of symptom management of acute pain and management of the behavioral symptoms of Alzheimer's disease patients. Dr. Leveck is currently on the board of governors of the NIH Clinical Center.

DIOMEDES LOGOTHETIS, Ph.D., is the dean of the Graduate School of Biological Sciences and acting director of the Medical Scientists Training Program at the Mount Sinai School of Medicine. He has been with the faculty of physiology and biophysics at Mount Sinai School of Medicine since 1993. His research, which is funded by the National Institutes of Health, National Science Foundation, and American Heart Association, is directed toward understanding in molecular terms how the activity of potassium ion channels is controlled by extracellular signals, such as hormones and neurotransmitters.

JOHN R. LUMPKIN, M.D., M.P.H., is senior vice president for health care at the Robert Wood Johnson Foundation (RWJF). Prior to joining RWJF, he was the first African American to hold the position of director of the Illinois Department of Public Health (IDPH). Dr. Lumpkin's career in public health began with his appointment in 1985 as associate director of IDPH's Office of Health Care Regulations, which oversees the licensing, inspection, and certification of healthcare facilities.

SHIRLEY M. MALCOM, Ph.D., is head of the Directorate for Education and Human Resources Programs of the American Association for the Advancement of Science. The directorate includes programs in education, activities for underrepresented groups, and public understanding of science and technology. From 1994 to 1998 Dr. Malcom served on the National Science Board, the policy-making body of the National Science Foundation, and from 1994 to 2001 she served on the President's Committee of Advisors on Science and Technology. In 2003 Dr. Malcom received the Public Welfare Medal of the National Academy of Sciences, the highest award given by the academy.

RICK A. MARTINEZ, M.D., is director of medical affairs for corporate community relations at Johnson & Johnson. Dr. Martinez has been director of CNS Medical Affairs at Janssen Pharmaceutical, associate director of the Janssen Research Foundation, and chief of geriatric psychopharmacology research with the National Institute of Mental Health. He is currently an issue expert for the Ad Council's Public Issues Committee.

ANGELA BARRON McBRIDE, R.N., Ph.D., is currently an Institute of Medicine nurse scholar. She is also dean emerita and distinguished professor at the Indiana University School of Nursing. Dr. McBride's research interests include the experience of parents, health concerns of women, and functional assessment of the seriously mentally ill. She is a member of the Institute of Medicine.

CRAIG McCLAIN, M.D., holds the University Distinguished Chair in Hepatology, serves as vice chair for research in the Department of Internal Medicine, and holds a graduate faculty appointment in pharmacology and toxicology at the University of Louisville. Prior to joining the faculty of the University of Louisville, he served as director of the NIH-funded General Clinical Research Center. Dr. McClain has more than 25 years of continuous federal funding, and his research focus is cytokines and liver disease.

NANCY E. REAME, Ph.D., R.N., is the Mary Dickey Lindsay Professor of Nursing and director of the DNSc Program at Columbia University. Previously she held the Rhetaugh G. Dumas Endowed Chair at the University of Michigan School of Nursing. She is an infertility nursing specialist, reproductive physiologist, and women's health researcher who conducts studies in the reproductive endocrinology of reproduction and menopause, and the bioethical aspects of assisted reproduction. Her current work is testing the theory that menopause starts in the brain, rather than in the ovary, where the gradual loss of eggs leads to a fall in estrogen. Her additional research includes programs looking at reproductive endocrinology, menstrual cycle, menopause, infertility, gender and health, and surrogate pregnancy. She initially studied nursing at Michigan State University and earned her master's and Ph.D. in maternity nursing and physiology at Wayne State. In 1980 she became a tenured professor at the University of Michigan. Dr. Reame is a member of the Institute of Medicine.

E. ALBERT REECE, M.D., Ph.D., M.B.A., is vice chancellor and dean of the College of Medicine at the University of Arkansas for Medical Sciences. Dr. Reece served on the faculty at Yale from 1982 to 1991 and was the Abraham Roth Professor and chair of the Department of Obstetrics, Gynecology, and Reproductive Sciences at the Temple University School of Medicine from 1991 to 2001. In addition, during this period he directed the Division of Maternal-Fetal Medicine and the Center for Fetal Diagnosis. His research focuses on diabetes in pregnancy, birth defects, and prenatal diagnosis. Dr. Reece is a member of the Institute of Medicine and of its Clinical Research Roundtable.

FRED SANFILIPPO, M.D., Ph.D., is the senior vice president for health sciences, dean of the College of Medicine and Public Health, and corporate executive officer of the Medical Center at Ohio State University. Dr. Sanfilippo was a member of the Duke University faculty from 1979 through 1992 and served as professor of pathology, surgery, and immunology, director of the Immunogenetics-Transplantation Laboratory, and chief of immunopathology. From 1993 to 2000 he was the Baxley Professor and Chair of Pathology and pathologist in chief at the Johns Hopkins Medical Institutions and served as director of the Johns Hopkins Medical Laboratories and director of research of the Johns Hopkins Comprehensive Transplant Center.

S. CLIFFORD SCHOLD JR., M.D., is associate vice chancellor for clinical research at the University of Pittsburgh Schools of the Health Sciences. Dr. Schold was formerly associated with the Duke Clinical Research Institute as director for neurosciences. He has also served as chair of the Department of Neurology at the University of Texas Southwestern Medical Center at Dallas. He is a neurologist with a subspecialty focus on neurooncology.

LARRY J. SHAPIRO, M.D., is the Spencer T. and Ann W. Olin Distinguished Professor and executive vice chancellor for medical affairs and dean of the School of Medicine at Washington University in St. Louis. Prior to joining the faculty at Washington University, Dr. Shapiro was the W. H. and Marie Wattis Distinguished Professor and chair of the Department of Pediatrics at the University of California San Francisco (UCSF), and chief of pediatric services at the UCSF Children's Hospital. Dr. Shapiro's research interests have included human molecular genetics and inborn errors of metabolism.

STEPHEN E. STRAUS, M.D., was appointed the first director of the National Center for Complementary and Alternative Medicine in October 1999. An internationally recognized expert in clinical research and clinical trials, Dr. Straus is also senior investigator in the Laboratory of Clinical Investigation at the National Institute of Allergy and Infectious Diseases (NIAID). He has extensive basic and clinical research experience related to many conditions for which there are alternative or complementary remedies, including chronic fatigue syndrome, Lyme disease, and HIV/AIDS. Dr. Straus's career at the National Institutes of Health began in 1979, when he joined NIAID as a senior investigator. Dr. Straus is board certified in internal medicine and infectious diseases.

NANCY S. SUNG, Ph.D., is a senior program officer with the Burroughs Wellcome Fund (BWF). She oversees BWF Interfaces in Science Programs, Innovation Awards in Functional Genomics (previous program), and Clinical Scientist Awards in Translational Research. Dr. Sung has also focused on building collaboration among other private foundations, government agencies, and professional societies who share BWF's interests in strengthening training and career pathways for researchers in the clinical research and physical or computational biology areas. Her research has focused on gene regulation in Epstein-Barr virus and its link to nasopharyngeal carcinoma, which is endemic in southern China. Prior to joining the BWF staff in 1997, Dr. Sung was a visiting fellow at the Chinese Academy of Preventive Medicine's Institute of Virology in Beijing.

NANCY FUGATE WOODS, R.N., Ph.D., is dean of the University of Washington School of Nursing. Dr. Woods is also founding director of the School of Nursing's internationally known Center for Women's Health Research and a former chair of the Department of Family and Child Nursing. She was previously the associate dean for research and has been a faculty member at the University of Washington since 1978. Dr. Woods has provided leadership since the 1970s in developing women's health as a field of study in nursing science. Her early research focused on the relationship between women's social environments and their health, emphasizing the health consequences of women's multiple roles and social supports.

NELDA WRAY, M.D., is chief research and development officer for the U.S. Department of Veterans Affairs, overseeing the VA research program. Before accepting her current position, Dr. Wray served as chief of general medicine at the Houston VA Medical Center and as professor and chief of health services research at the Baylor College of Medicine in Houston. In 1999 she became the second person to receive the VA Under Secretary Award for Outstanding Achievement in Health Service Research. Dr. Wray is board certified in internal medicine and pulmonary medicine.

JOHN YATES, M.D., is president of Takeda Global Research and Development Center Inc. Previously, he served as vice president, Medical and Scientific Affairs (MEDSA) of Merck & Co. Prior to joining MEDSA, Dr. Yates was vice president of clinical development in the U.S. Human Health Division of Merck. He also worked with the then-nascent osteoporosis clinical research team at Merck and subsequently led that team in the development of bisphosphonate alendronate (Fosamax) for treatment and prevention of postmenopausal osteoporosis.

Appendix B

Workshop Guests

Debra Aronson
Federation of American Societies for Experimental Biology

Catherine Baase
Dow Chemical Company

Claudia R. Baquet
University of Maryland School of Medicine

Diane Bernal
National Eye Institute, National Institutes of Health (NIH)

Queta Bond
Burroughs Wellcome Fund

Robert Bonow
Northwestern University Feinberg School of Medicine

Beth Bowers
National Institute of Mental Health, NIH

Barbara Bowman
Centers for Disease Control and Prevention

David Burnaska
U.S. Department of Veterans Affairs

Scott Campbell
American Diabetes Association

Veronica Catanese
New York University School of Medicine

Francis Chesley
Agency for Healthcare Research and Quality

Michelle Cissell
Juvenile Diabetes Research Foundation

Lois Colburn
Association of American Medical Colleges

Elaine Collier
National Center for Research Resources, NIH

Claire Cornell
National Association of Community Health Centers

Thomas Crist
Alliance for Academic Internal Medicine

William F. Crowley
Massachusetts General Hospital

Joan Davis
National Institute of Child Health and Human Development, NIH

Claude Desjardins
University of Illinois

Adrian Dobs
Johns Hopkins University School of Medicine

APPENDIX B

Jessica Donze
American Dietetic Association

Andrew Fishleder
Cleveland Clinic Foundation

Maryrose Franko
Howard Hughes Medical Institute

William R. Galey
Howard Hughes Medical Institute

Elaine Gallin
Doris Duke Charitable Foundation

Myron Genel
Yale University School of Medicine

Kenneth Getz
CenterWatch

Marian Girardi
American Academy of Otolaryngology

Maureen Hannley
American Academy of Otolaryngology

Anthony Hayward
National Center for Research Resources, NIH

Carlton Hornung
University of Louisville

Grant Huang
U.S. Department of Veterans Affairs

Bonnie Jennings
American Academy of Nursing

Stephen B. Johnson
Columbia University

William N. Kelley
University of Pennsylvania School of Medicine

Mahin Khatami
Molecular Technologies

Michael Klag
Johns Hopkins University, School of Medicine

Dushanka Kleinman
National Institute of Dental and Craniofacial Research, NIH

Allan M. Korn
Blue Cross/Blue Shield Association

David Korn
Association of American Medical Colleges

Theodore Kotchen
Medical College of Wisconsin

Steven Krosnick
National Cancer Institute, NIH

Joel Kupersmith
Institute of Medicine

Jennette Lawrence
American College of Surgeons

Mary D. Leveck
National Institute of Nursing Research, NIH

John R. Lumpkin
Robert Wood Johnson Foundation

Maria Majewska
National Institute on Drug Abuse, NIH

Shirley M. Malcom
American Association for the Advancement of Science

Rick A. Martinez
Johnson & Johnson

Angela Barron McBride
Indiana University School of Nursing

Kenneth W. Miller
American Association of Colleges of Pharmacy

Nancy E. Miller
Office of the Director, NIH

Jay Moskowitz
Penn State University

Claudia Moy
National Institute of Neurological Disorders and Stroke, NIH

Esther Myers
American Dietetic Association

Asua Ofosu
American Thoracic Society

Delores Parron
Office of the Director, NIH

Nancy E. Reame
Columbia University School of Nursing

E. Albert Reece
University of Arkansas College of Medicine

David L. Rimoin
Cedars-Sinai Medical Center

Ann Rose
VICRO

Michael Sayre
National Center for Research Resources, NIH

Bernard Schwetz
Office of Public Health and Science

Louis Sherwood
MESDA, LCC
Merck & Co. (retired)

Lawrence Shulman
National Institute of Arthritis and Musculoskeletal and Skin Diseases, NIH

Harold Slavkin
University of Southern California School of Dentistry

Paul Smedberg
American Society of Nephrology

Stephen Sonstein
Eastern Michigan University

Robert Star
National Institute of Diabetes and Digestive and Kidney Diseases, NIH

Stephen E. Straus
National Center for Complementary and Alternative Medicine, NIH

Carolyn Strete
National Cancer Institute, NIH

Nancy S. Sung
Burroughs Wellcome Fund

Crispin Taylor
Science, American Association for the Advancement of Science

Julie Taylor
American Society of Clinical Oncology

Hugh Tilson
University of North Carolina School of Public Health

Anton-Lewis Usala
East Carolina University

Donald Vereen
National Institute on Drug Abuse, NIH

Frances Visco
National Breast Cancer Coalition

James Voytuk
National Research Council

Jacqueline Whitted
National Cancer Institute, NIH

Jennifer Wilson
Annals of Internal Medicine

Appendix C

Workshop Agenda

October 16, 2003
National Academy of Sciences Auditorium
Washington, D.C.

9:00 a.m.   *Assessment of Progress*

           William N. Kelley, M.D.
           Professor of Medicine, Biochemistry, and Biophysics
           University of Pennsylvania School of Medicine

           William Crowley, M.D.
           Director of Clinical Research
           Massachusetts General Hospital

10:45 a.m.   Break

11:00 a.m.   *Emerging Recruitment Issues and Workforce Needs*

           Nelda Wray, M.D.
           Chief Research and Development Officer
           U.S. Department of Veterans Affairs

Rick A. Martinez, M.D.
Director, Medical Affairs
Johnson & Johnson

John Yates, M.D.
Vice President, Medical and Scientific Affairs
Merck & Co.

12:00 p.m. Lunch

1:00 p.m. *Keynote Address: Re-engineering the Clinical Research Enterprise*

Introduction
Vivian W. Pinn, M.D.
Director
Office of Research on Women's Health

Stephen E. Straus, M.D.
Director
National Center for Complementary and Alternative Medicine

2:00 p.m. *Interdisciplinary Research and Emerging New Skill Sets: The Role of Nurses in Clinical Research*

Session Moderator
Nancy Reame, R.N., Ph.D.
Professor of Nursing
University of Michigan School of Nursing

Panel:
Mary D. Leveck, R.N., Ph.D.
Deputy Director
National Institute of Nursing Research

Angela Barron McBride, R.N., Ph.D.
Distinguished Professor and Dean Emerita
Indiana University School of Nursing

Nancy F. Woods, R.N., Ph.D.
Dean, School of Nursing
University of Washington

3:45 p.m. Break

4:00 p.m. *Strengthening Career Pathways to Promote Diversity*

Claudia R. Baquet, M.D.
Associate Dean for Policy and Planning
University of Maryland, School of Medicine

William R. Galey, Ph.D.
Director, Graduate Science Education Program
Howard Hughes Medical Institute

Shirley M. Malcom, Ph.D.
Head, Directorate for Education and Human Resources
American Association for the Advancement of Science

Lois Colburn
Assistant Vice President, Division of Community and Minority Programs
Association of American Medical Colleges

5:30 p.m. Adjourn

October 17, 2003
National Academy of Sciences Auditorium
Washington, D.C.

9:00 a.m. *The Role of "Group Therapy" in Enhancing the Clinical Research Enterprise*

E. Albert Reece M.D., Ph.D., M.B.A.
Vice Chancellor and Dean
University of Arkansas College of Medicine

9:45 a.m.	*Roundtable Discussion: Since translational research often involves more collaborative research, how will the academic system acknowledge collaborative research in its tenure process?* *Discussion leader* E. Albert Reece M.D., Ph.D., M.B.A.
11:00 a.m.	Break
11:15 a.m.	*Roundtable Discussion: How should the academic community make clinical/translational research a priority and regard it as a completion of basic science findings?* *Discussion leader* E. Albert Reece M.D., Ph.D., M.B.A.
12:30 p.m.	Lunch
1:15 p.m.	*Roundtable Discussion: How does the academic community view the translational blocks identified by members of the Clinical Research Roundtable?* *Opening presentation* Nancy S. Sung, Ph.D. Program Officer Burroughs Wellcome Fund
2:30 p.m.	*Roundtable Discussion: How can the private sector most effectively collaborate on training issues with medical schools?* *Opening presentation* Rick A. Martinez, M.D. Director, Medical Affairs Johnson & Johnson John Yates, M.D. Vice President, Medical and Scientific Affairs Merck & Co.

3:45 p.m.   Break

4:00 p.m.   *Roundtable Discussion: Clinical research training, diversity, and future direction at the Robert Wood Johnson Foundation*

Opening Presentation
John R. Lumpkin, M.D., M.P.H.
Senior Vice President for Health Care
Robert Wood Johnson Foundation

5:00 p.m.   Adjourn

Appendix D

Public Mechanisms for Clinical Research Training: Examples of Minority Research Training Programs

Table begins on next page.

Name of Program	Focus
NIH	
MARC/U*STAR (T34)	To increase the number of underrepresented minority researchers in biomedical research sciences

URL: http://grants1.nih.gov/grants/guide/pa-files/PAR-02-033.html.

Ruth L. Kirschstein National Research Service Award Predoctoral Fellowship for Minority Students (F31)	To enhance the racial and ethnic diversity of the biomedical, behavioral, and health services research labor force in the United States

URL: http://grants.nih.gov/grants/guide/pa-files/PA-00-069.html

Ruth L. Kirschstein NRSA Program for NIGMS MARC Predoctoral Fellowships (F31)	To encourage students from minority groups underrepresented in the biomedical and behavioral sciences to seek graduate degrees

URL: http://grants.nih.gov/grants/guide/pa-files/PAR-03-114.html

Mental Health Dissertation Research Grants to Increase Diversity in the Mental Health Research Arena (R36)	To enable doctoral candidates from racial and ethnic groups underrepresented in biomedical and behavioral science to pursue research careers in areas relevant to the research mission of NIMH

URL: http://grants.nih.gov/grants/guide/pa-files/PAR-03-110.html

Requirements	No. of Awards	Length of Program	Total Program Support
Undergraduate junior and senior honor students majoring in the biomedical sciences with an expressed interest in a career in biomedical research and intentions to pursue graduate education leading to a Ph.D., M.D.-Ph.D., or other combined professional degree and Ph.D.	54 universities, (supporting about 700 students)	Two years	**$1.4 million** FY 2002
Up to five years of support for research training leading to the Ph.D. or equivalent research degree; the combined M.D.-Ph.D. degree or other combined professional degree and research doctoral degree in the biomedical or behavioral sciences or health services research	389 individual fellowships in 2002	Up to five years	**$25,796,400** FY 2002
For selected students who are graduates of the MARC undergraduate research training programs, up to five years of support for research training leading to the Ph.D., M.D.-Ph.D., or other combined professional degree and Ph.D. in the biomedical or behavioral sciences	137 individual fellowships (FY2002) 159 individual fellowships (FY 2003)	Up to five years	**$4,016,000** FY 2002 **$4,866,000** FY 2003
Doctoral candidates in an area or discipline relevant to the mission of NIMH and conducting dissertation research	4 individual fellowships (FY 2002) 3 (FY 2003) 4 (FY 2004)	Up to two years	**$107,000** FY 2002 **$72,000** FY 2003 **$120,000** FY 2004

Name of Program	Focus
NIH	
Ruth L. Kirschstein National Research Service Award: NHLBI Minority Institutional Training Grants (T32)	To support training of graduate and health professional students and individuals in postdoctoral training at minority schools having the potential to develop meritorious training programs in cardiovascular, pulmonary, hematologic, and sleep disorders
URL: http://grants1.nih.gov/grants/guide/rfa-files/RFA-HL-04-027.html	
Ruth L. Kirschstein National Research Service Award: NHLBI Short-Term Training for Minority Students Program (T35)	To provide short-term research support to underrepresented minority undergraduate and graduate students and students in health-professional schools to provide them with career opportunities in research
URL: http://grants1.nih.gov/grants/guide/rfa-files/RFA-HL-03-014.html	
Loan Repayment Program for Health Disparities Research	To provide an incentive for health professionals to engage in basic, translational, or behavioral research directly relevant to health disparities research
URL: http://grants.nih.gov/grants/guide/notice-files/NOT-MD-04-001.html	
Loan Repayment Program for Individuals from Disadvantaged Backgrounds	To provide an incentive for health professionals from disadvantaged backgrounds to conduct translational research and promote the development of research programs that reflect the variety of issues and problems associated with disparities in health status
URL: http://ncmhd.nih.gov/our_programs/loan/index.asp	

Requirements	No. of Awards	Length of Program	Total Program Support
Predoctoral students must be training at the post-baccalaureate level and enrolled in a program leading to a Ph.D. or be a health-professional student or individual in postgraduate clinical training, and wish to interrupt their studies for a year or more to engage in full-time research training Postdoctoral trainees must have received a Ph.D., M.D., D.D.S., or comparable degree	5 institutional awards, supporting 32 individuals	Up to five years	**$734,080** (FY 2004)
Trainees must have successfully completed at least one undergraduate year or have successfully completed one semester at a school of medicine (or other health profession) or graduate program	8 institutional awards	Two to three months	**$500,000** FY 2004
Qualified health professionals who contractually agree to conduct qualified minority health disparities research or other health disparities research for 50 percent of time, or not less than 20 hours per week	121 individual awards	Two years	**$2,036,000** FY 2003
An "individual from a disadvantaged background" is one who comes from a family with an annual income below a level based on low-income thresholds according to family size published by the U.S. Census Bureau	33 individual awards	Two years	**$1,332,000** FY 2003

Name of Program	Focus
NIH	
Project EXPORT	To build research capacity at designated institutions enrolling a significant number of students from health disparity populations and to promote participation and training in biomedical and behavioral program populations

URL: http://ncmhd.nih.gov/our_programs/project_export_awards/PrjExpFY03Awards.asp

Agency for Healthcare Research and Quality (AHRQ)	
AHRQ Minority Research Infrastructure Support Program (M-RISP)	To increase the capacity of institutions that serve racial or ethnic minorities in Hawaii, Tennessee, and Texas

URL: http://grants1.nih.gov/grants/guide/pa-files/PAR-01-001.html

Requirements	No. of Awards	Length of Program	Total Program Support
Designated institutions must have a significant number of members of minority health disparity populations or other health disparity populations enrolled as students; have been effective at assisting such students' research on health disparity to complete the education or training and receive the degree involved; have made significant efforts to recruit minority students to enroll in and graduate from the institution; have made significant recruitment efforts to increase the number of minority or other members of health disparity populations serving in faculty or administrative positions at the institution	33 institutional awards	Three years	**$32 million** FY 2003
Applications may be submitted by domestic public and private colleges, for their faculty to conduct rigorous health services research, as well as nonprofit domestic organizations for first-year funding (e.g., hospitals, laboratories, $150,000-$522,022); units of public agencies of state or local governments, eligible agencies of the federal government, or other institutions conducting health services research	9 institutional awards	Three years	**$1,179,919** FY 2003

Name of Program	Focus

Health Resources and Services Administration (HRSA)

HRSA Centers of Excellence HRSA-04-004	To assist eligible schools in supporting programs of excellence in health professions education for underrepresented minority individuals

URL: http://www.hrsa.gov/grants/preview/professions.htm#hrsa04004

HRSA-04-011 Nursing Workforce Diversity	To increase nursing education opportunities for individuals from disadvantaged backgrounds (including racial and ethnic minorities among registered nurses) by providing student scholarships or stipends, pre-entry preparation, and retention activities

URL: http://www.hrsa.gov/grants/preview/professions.htm

HRSA-04-009 Health Careers Opportunity Program (HCOP)	To assist individuals from disadvantaged backgrounds to undertake education to enter a health profession

URL: http://www.hrsa.gov/grants/preview/professions.htm

Requirements	No. of Awards	Length of Program	Total Program Support
Eligible applicants are accredited schools of allopathic medicine, osteopathic medicine, dentistry, pharmacy, graduate programs in behavioral or mental health, or other public and nonprofit health or educational entities, including faith-based organizations and community-based organizations, that meet the requirements of section 736(c) of the Public Health Service Act	10 institutional awards	3 years	**$6,118,398** FY 2004
Schools of nursing, nursing centers, academic health centers, state or local governments, underrepresented American Indian tribes or tribal organizations, and other public or private nonprofit entities	39 institutional awards	3 years	**$11,396,900** FY 2004
Eligible applicants include schools of medicine, osteopathic medicine, public health, dentistry, veterinary medicine, optometry, pharmacy, allied health, chiropractic, podiatric medicine, public or non-profit private schools that offer graduate programs in behavioral and mental health, programs for the training of physician assistants, and other public or private nonprofit health or educational entities, including faith-based and community-based organizations	35 individuals	3 years	**$14,152,621** FY 2004

Appendix E

Public Mechanisms for Clinical Research Training

Name of Program	Focus
NIH	
Mentored Clinical Scientist Development Award (K08)	To support the development of outstanding clinician research scientists—three to five years; 75 percent
URL: http://grants1.nih.gov/grants/guide/pa-files/PA-00-003.html	
Mentored Clinical Scientist Development Program Award (K12)	To support an institution for the development of independent translational and nonprofit organizations, scientists—five years; 75 percent effort
URL: http://grants1.nih.gov/training/careerdevelopmentawards.htm	
Mentored Patient-Oriented Research Career Development Award (K23)	To support the career development of investigators who have made a commitment to focus their research endeavors on patient-oriented research—three to five years; 75 percent commitment
URL: http://grants1.nih.gov/grants/guide/pa-files/PA-00-004.html	
Midcareer Investigator Award (K24)	To provide support for patient-oriented research; to allow them protected time to devote to patient-oriented research and to act as mentors for beginning translational investigators—three to five years; 25-50 percent effort
URL: http://grants1.nih.gov/grants/guide/pa-files/PA-00-005.html	

Requirements	No. of Awards	Length of Program	Total Program Support
M.D., D.D.S., Pharm.D., Ph.D. in nursing	1,161	Three to five years	**$143,731,640** FY 2002
Institutions: public and private; clinicians: M.D., D.O., doctorally prepared registered nurses	1,073	Five years	**$55,053,672** FY 2002
M.D., D.D.S., Pharm.D., Ph.D in translational nursing	80 (664 for 2002)	Three to five years	**$88,296,497** FY 2002
Clinicians: M.D., D.D.S., Pharm.D., Ph.D. in translational nursing	60-80 (261 for 2002)	Three to five years	**$28,747,290** FY 2002

Name of Program	Focus
NIH	
Clinical Research Curriculum Award (K30)	To attract talented individuals to the challenges of translational research to and to provide them with the critical skills that are needed to develop hypotheses and conduct sound regulatory research—five years; up to $200,000 per year
URL: http://grants1.nih.gov/training/K30.htm	
Loan Repayment Program (LRP)	To attract health professionals to clinical research
URL: http://www.lrp.nih.gov/about/extramural/intro.htm#clinical	
National Research Service Award (NRSA): Physicians and Doctorally Prepared Nurses	
NRSA T32	Eligible institutions to develop or enhance research training opportunities for individuals, selected by the institution who are training for careers in specified areas of biomedical, behavioral, and translational research
URL: http://grants1.nih.gov/grants/guide/pa-files/PA-02-109.html	
NRSA T35	To eligible institutions to develop or enhance research training for individuals interested in career opportunities
URL: http://grants1.nih.gov/grants/guide/notice-files/not98-027.html	

Requirements	No. of Awards	Length of Program	Total Program Support
Awarded to institutions; is a trans-NIH program. Formal coursework includes design of translational research projects, hypothesis development, biostatistics, epidemiology, legal and ethical issues related to translational research	59	Five years	**$12,094,341** FY 2002
Qualified health professionals who contractually agree to conduct qualified translational research for 50 percent of time, or not less than 20 hours per week, for a two-year consecutive period	1,200	Two years	**$63.3 million** FY 2003
Ph.D., D.D.S., M.D., or a comparable doctoral degree	17,000 (total no. of graduate students); 386 in 2002	Five years	**$497,424,833** FY 2002
This can be used to support other types of predoctoral and postdoctoral training in focused, often emerging, or scientific areas	35 trainees per budget period	Five years	**$7,959,251** FY 2002

Name of Program	Focus

Health Resources and Services Administration (HRSA): Doctorally Prepared Nurses

HRSA-04-010 Advanced Education Nursing	Awarded to institutions for projects that support the enhancement of appropriate legislative purpose

***URL:** http://www.hrsa.gov/grants/preview/professions.htm*

HRSA-04-012 Advanced Education Nursing Traineeships	Eligible institutions meet the cost of traineeships for individuals in advanced nursing education programs

***URL:** http://www.hrsa.gov/grants/preview/professions.htm*

HRSA-04-011 Nursing Workforce Diversity	To increase nursing education opportunities for individuals from disadvantaged backgrounds (including racial and ethnic minorities underrepresented among registered nurses) by providing student scholarships or stipends, pre-entry preparation, and retention activities

***URL:** http://www.hrsa.gov/grants/preview/professions.htm*

Department of Veterans Affairs (VA): Physicians

VA Mentored Research Training Programs	To provide mentored research health services training for various stages of a clinician's career development

***URL:** http://www.appc1.va.gov/resdev/ps/psmr/mentored_research.htm*

Requirements	No. of Awards	Length of Program	Total Program Support
Schools of nursing, academic health centers, public or private nonprofit entities, and for-profit entities capable of carrying out advanced nursing education and practice	8	Three years	**$2,091,892** FY 2004
Schools of nursing, academic health centers, other appropriate public or private nonprofit entities, and for-profit entities capable of carrying out the legislative purpose	335	One year	**$4,800,000** FY 2004
Schools of nursing, nursing centers, academic health centers, state or local governments, American Indian tribes or tribal organizations, other public or private nonprofit entities	39	Three years	**$11,396,00** FY 2004
Programs are for the associate investigator, career development, advanced career research, and merit review entry program; M.D., D.D.S., Ph.D.	81	One to five years	**$389,978** 1998-1999

Appendix F

Examples of Pharmaceutical Company Training Programs

Allergan Inc.

1. Name of Program: AGS Allergan Grant—2004 Clinician-Scientist Fellowships in Glaucoma
Population: Completed within the last five years at least one full year of fellowship training in glaucoma
Program Description:
- To encourage the development of new clinician-scientists in glaucoma

Length of Award: One year
Amount Awarded: $35,000
Application Information: http://www.glaucomaweb.org/award2004.html

Amgen

1. Name of Program: Amgen (Washington) Postdoctoral Program
Population: Fellowship candidates must have a Ph.D. and/or M.D. and a track record of research accomplishment.
Program Description:
- Work closely with the fellow to identify mentors and projects that match his or her research interests
- Present research findings at both internal and external meetings

Length of Award: Three years
Application Information: http://www.amgen.com/career/PostDoc/index.html

2. Name of Program: American Association for Cancer Research (AACR)–Amgen, Inc., Fellowship in Clinical or Translational Cancer Research
Population: Candidate must have been a fellow for at least two years (24 months) but not more than four years (48 months) prior to the beginning of the award year.
Program Description:
- To foster basic, translational, clinical, and prevention research by scientists at the beginning of their career in the cancer field

Length of Award: One year
Amount Awarded: $35,000
Application Information: http://www.aacr.org/1605.asp

Aventis

1. Name of Program: Medical Information Fellowship Program
Population: Pharm.D.
Program Description:
- To provide efficient and unbiased medical information on Aventis Pharmaceuticals products to healthcare professionals, consumers, and internal associates
- Complete a fellowship research project to be presented at a national conference

Length of Award: One year
Application Information: http://pharmacy.rutgers.edu/fellows

2. Name of Program: Global Drug Information Residency Program
Population: Pharm.D. or B.S. in pharmacy with a general pharmacy practice residency or two years, clinical working experience
Program Description:
- Provide accurate and unbiased global drug and medical information efficiently

Length of Award: One year
Application Information: http://www.aventispharma-us.com/residencies/ApplyNow.jsp

Baxter

1. Name of Program: National Hemophilia Foundation (NHF) Clinical Fellowship Program
Population: Institutions must have well-established hemophilia or thrombophilia treatment centers with qualified clinical and research faculty. Must have a medical degree, prefellowship clinical training
Program Description:
- To increase the number of clinicians who are dedicated to providing care to patients with bleeding disorders, such as hemophilia and to prepare fellows for academic careers

Length of Award: Ongoing for five years with funding awarded to institutions for up to two years
Amount Awarded: Awards to institutions will be up to $100,000 per fellow per year
Application Information: http://www.hemophilia.org/research/RFA_clinicalfellowship.pdf

Bristol-Myers Squibb

1. Name of Program: Fellowship Program in Academic Medicine for Minority Students
Population: First- through third-year students. U.S. citizens who are African American, mainland Puerto Rican, Mexican American, or American Indian (Alaska Native, Native Hawaiian); M.D.-degree-granting medical school in the United States
Program Description:
- To address a simple but significant statistic: minorities are profoundly underrepresented in the field of academic medicine

Length of Award: 8-12 weeks
Amount Awarded: $6,000
Application Information: http://www.bms.com/sr/philanthropy/data/fellow2003.pdf

2. Name of Program: Freedom to Discover Grants and Awards
Population: Institutions and principal investigators: cancer grant recipients, cardiovascular grant recipients, infectious diseases grant recipients, metabolic grant recipients, neuroscience grant recipients, and nutrition grant recipients
Program Description:
- To provide relief and even cures from devastating and debilitating illnesses to millions of people around the world

Length of Award: Five years
Amount Awarded: $500,000 unrestricted research grant and a distinguished achievement award of $50,000 to an individual researcher
Application Information: http://www.bms.com/sr/grants/data/factsh.doc

Eli Lilly

1. Name of Program: Drug Information Residency
Population: Pharm.D. students
Program Objective:
- *Pharmaceutical Industry Setting:* To receive balanced instruction in service, education, and research
- *Institutional Setting:* To gain important experience in the provision of drug information within the acute care setting, including participation in the Pharmacy and Therapeutics Committee and the Institutional Review Board

Length of Award: One year
Application Information: http://www.lilly.com/careers/campuszone/

2. Name of Program: Visiting Scientist Program
Population: Pharm.D., Ph.D., M.D., or master's degree
Program Objective:
- To fully develop individuals into effective, influential professionals knowledgeable about the drug development process and the pharmaceutical industry

Length of Award: One year
Amount Rewarded: $32,000 per year
Application Information: http://www.lilly.com/careers/campuszone/

3. Name of Program: Damon Runyon-Lilly Clinical Investigator Award
Population: M.D. or M.D.-Ph.D. degree(s); applicants may apply during the final year of their subspecialty training or within the first four years of their assistant professorship appointment
Program Description:
- To increase the number of physicians capable of moving seamlessly between the laboratory and the patient's bedside in search of breakthrough treatments

Length of Award: Five years
Amount Awarded: Year 1: $30,000; Year 2: $30,000; Year 3: $20,000; Year 4: $15,000; Year 5: $0.
The foundation will also retire up to $100,000 of any medical school debt still owed by the awardee
Application Information: http://www.drcrf.org/apClinical.html

GlaxoSmithKline

1. Name of Program: Drug Development/Clinical Research-Philadelphia College of Pharmacy
Population: Pharm.D. or Ph.D.
Program Description:
- Includes development of research skills to permit independent investigation of cardiovascular problems in humans through a hands-on approach

Length of Award: Two years
Amount Awarded: $37,000 for the first year; $38,000 for the second year
Application Information: http://www.gsk.com/careers/us-university/university_us_residencies.htm

2. Name of Program: Global Health and Clinical Outcomes Fellowship-Thomas Jefferson University
Population: Pharm.D., Ph.D., or M.D.
Program Description:
- To build the fellow's prior knowledge and skills with training in economic and health services research methodologies

Length of Award: Two years
Application Information: http://www.gsk.com/careers/us-university/university_us_residencies.htm

3. Name of Program: Pharmacoeconomics/Health Outcomes Fellowship-University of North Carolina (UNC) at Chapel Hill
Population: Applicants must have completed all coursework prior to starting the fellowship and be enrolled in either the Health Policy and Administration (HPAA) Ph.D. program or the Pharmaceutical Policy and Evaluative Sciences (PPES) Ph.D. program at UNC at Chapel Hill.
Program Description:
- To prepare fellows for careers in health economics, health outcomes research, or pharmacoeconomics

Length of Award: Two years
Amount Awarded: $32,000 per year
Application Information: http://www.gsk.com/careers/us-university/university_us_residencies.htm

4. Name of Program: Pharmacokinetics/Pharmacodynamics Fellowship, University of North Carolina at Chapel Hill
Population: Pharm.D., Ph.D., or M.D. with advanced training or experience in pharmacokinetics
Program Description:
- To provide knowledge and experience in clinical pharmacokinetic/dynamic study design and related drug research methodology D1-3

Application Information: http://www.gsk.com/careers/us-university/university_us_residencies.htm

5. Name of Program: Drug Development/Clinical Research, University of North Carolina at Chapel Hill
Population: Pharm.D.
Program Description:
- To provide knowledge and experience in study design and methodology, analytical techniques, proper conduct of clinical drug trials, and exposure to ethical, legal, and regulatory issues in research involving investigational and marketed drugs

Length of Award: Two years
Application Information: http://www.gsk.com/careers/us-university/university_us_residencies.htm

6. Name of Program: Specialty Residency in Medical Information—GlaxoSmithKline (GSK) and Duke University Medical Center (DUMC)
Population: Pharm.D.
Program Description:
- Participate in numerous activities that a medical information specialist encounters

Length of Award: 12 months
Application Information: http://www.gsk.com/careers/us-university/university_us_residencies.htm

7. Name of Program: Postdoctoral Fellowship in Oncology/Clinical Pharmacology, Duke University Medical Center, Comprehensive Cancer Center
Population: M.D. degree and licensed to practice medicine in North Carolina; completed an accredited residency program and be in a fellowship program in adult or pediatric hematology-oncology
Program Description:
- To prepare physicians for careers in pharmaceutical research and development, in academia, in clinical research, or in a government agency that deals with drugs or therapeutics

Length of Award: Two years
Amount Awarded: PGY Level 4 (typical entering level for fellows): $44,952 per year
Application Information: http://www.gsk.com/careers/us-university/university_us_residencies.htm

Johnson & Johnson

1. Name of Program: The Woodrow Wilson-Johnson & Johnson Dissertation Grants
Population: Students in doctoral programs such as nursing, public health, anthropology, history, sociology, psychology, and social work at graduate schools in the United States
Program Description:
- To encourage original and significant research on issues related to women's health

Length of Award: Two years
Amount Awarded: $3,000 to be used for expenses connected with the dissertation
Application Information: http://www.woodrow.org/womens-studies/purpose.html

2. Name of Program: Johnson & Johnson Co-op
Population: Medical students
Program Description:
- To apply academic knowledge in business settings

Length of Award: *Intern*: three months full time; *Co-op*: six months full time
Application Information: http://www.jnj.com/careers/intcoop.html

Merck

1. Name of Program: 2004 Merck/AFAR (American Federation for Aging Research) Junior Investigator Awards in Geriatric Clinical Pharmacology
Population: Board certified or eligible in a primary specialty by July 1, 2004; be within four years of having completed postdoctoral or fellowship training
Program Description:
- To develop a cadre of physicians with a command of the emerging field of geriatric clinical pharmacology

Length of Award: Two years
Amount Awarded: $120,000 over two years
Application Information: http://www.merck.com/about/cr/policies_performance/social/focus.html

2. Name of Program: American Gastroenterological Association (AGA)–Merck Clinical Research Career Development Award
Population: Junior faculty members performing clinical research
Program Description:
- To provide support for research so the awardee can develop an independent and productive career as a clinical investigator in any area of gastroenterology or hepatology

Length of Award: Two years
Amount Awarded: $25,000 per year
Application Information: http://www.fdhn.org/html/pdf/descriptions/MerckAwardDescription.pdf

3. Name of Program: United Negro College Fund (UNCF)–Merck Science Initiative
Population: African American students pursuing studies and careers in chemistry and the life sciences; undergraduate, graduate, and postdoctoral levels, administered by the College Fund/UNCF
Program Description:
- To expand the pool of world-class African American biomedical scientists and achieve national economic competitiveness and social diversity

Length of Award: *Graduate Science Research Dissertation Fellowships*: one to two years of fellowship tenure; *Postdoctoral Science Research Fellowships*: 12 years
Amount Awarded: *Graduate Science Research Dissertation Fellowships*: $20,000 per year; *Postdoctoral Science Research Fellowships*: $35,000 per year
Application Information: http://www.uncf.org/Merck/

4. Name of Program: American College of Cardiology (ACC)–Merck Adult Cardiology Research Fellowship Awards
Population: American College of Cardiology selects awardees, known as "Merck Fellows of the ACC"
Program Description:
- Receive advanced training in adult cardiology

Length of Award: One year
Amount Awarded: $40,000
Application Information: http://www.acc.org/about/award/awardopps.htm#fellowship

5. Name of Program: Merck Sharp & Dohme International Fellowships in Clinical Pharmacology
Population: Residents and citizens of countries other than the United States, graduates of accredited medical schools and licensed to practice medicine in their home countries
Program Description:
- To provide training for individuals from countries outside the United States so that they can become qualified to teach and conduct research in medical schools and hospitals around the world

Length of Award: One to two years
Amount Awarded: $30,000 per year
Application Information: http://www.merck.com/about/cr/policies_performance/social/focus.html#ClinicalPharmacology

Pfizer

1. Name of Program: Medical and Academic Partnerships (MAP) grants and awards
Population: Physician-scientists
Program Description:
- To support medical innovation in a wide range of disciplines
- Awards available include Fellowships, Scholar Grants, Visiting Professorships, the Clinical Research Training Program (CRTP)

Length of Award: *Fellowships*: two to three years; *Scholar Grants*: two to three years; *Visiting Professorships*: three days; and *CRTP*: one year
Amount Awarded: *Fellowships*: $65,000 per year; *Scholars Grants*: $65,000 per year; *Visiting Professorships*: $7,500 per institution; and *CRTP*: $27,100 per year
Application Information: http://www.physicianscientist.com

2. Name of Program: Chest Foundation Clinical Research Trainee Awards 2004
Population: Physicians enrolled in a U.S. or Canadian subspecialty training program in the following disciplines: allergy or immunology, cardiac electrophysiology, critical care anesthesiology, critical care intensive care, critical care medicine, infectious disease, cardiology, pediatric critical care, pediatric pulmonary disease, pulmonary disease, surgical critical care, or thoracic surgery
Program Description:
- To support clinical research and not basic or bench-level research in asthma, chronic obstructive pulmonary disease, pulmonary fibrosis, or women's health; applicants may apply for only one type of clinical research award

Length of Award: One year
Amount Awarded: $10,000
Application Information: http://www.chestfoundation.org/awards/clinical/index.php

3. Name of Program: Pfizer Ophthalmics Research Fellowship in Glaucoma—Clinician-Scientist Fellowships in Glaucoma
Population: Completed within the last five years at least one full year of fellowship training in glaucoma
Program Description:
- To encourage the development of new clinician-scientists in glaucoma

Length of Award: One year
Amount Awarded: $40,000
Application Information: http://www.glaucomaweb.org/associations/5224/files/Announcement%202006.pdf

PhRMA: America's Pharmaceutical Research Companies

1. Name of Program: Awards in Health Outcomes—Predoctoral Fellowships, Postdoctoral Fellowships, Research Starter Grants
Population: *Predoctoral:* student's Ph.D. doctoral program after coursework has been

completed and the remaining training activity is the student's research project; *Postdoctoral:* graduates from Pharm.D., M.D., and Ph.D. programs; *Research Starter:* applicants must be appointed to an entry-level tenure track or equivalent permanent position in a department or unit responsible for health outcomes research activities as part of its core mission

Program Description:
- *Predoctoral:* To provide some assistance in this training sequence
- *Postdoctoral:* To support postdoctoral career development activities of individuals
- *Research Starter:* To offer financial support to individuals beginning their independent research careers

Length of Award: *Predoctoral:* two years; *Postdoctoral:* two years; *Research Starter:* two years

Amount Awarded *Predoctoral:* $20,000 per year; *Postdoctoral:* $40,000 per year; *Research Starter:* $30,000 per year

Application Information: http://www.phrmafoundation.org to download an application and for the specific requirements

2. Name of Program: Awards in Pharmacology/Toxicology—Predoctoral Fellowships, Postdoctoral Fellowships, Research Starter Grants

Population: *Predoctoral:* full-time, in-residence Ph.D. candidates in the fields of pharmacology or toxicology who are enrolled in U.S. schools of medicine, pharmacy, dentistry, or veterinary medicine; *Postdoctoral:* (1) hold a Ph.D. degree or appropriate terminal research doctorate in a field of study logically or functionally related to the proposed postdoctoral activities or (2) expect to receive the Ph.D. before activating the award; *Research Starter:* instructor or assistant professor and investigators at the doctoral level with equivalent positions, providing their proposed research is neither directly nor indirectly subsidized to any significant degree by a competitive extramural grant

Program Description:
- To support career development activities of scientists prepared to engage in research that integrates information on molecular or cellular mechanisms of action with information on the effects of an agent observed

Length of Award: *Predoctoral:* two years; *Postdoctoral:* two years; *Research Starter:* two years

Amount Awarded *Predoctoral:* $20,000 per year; *Postdoctoral:* $40,000 per year; *Research Starter:* $30,000 per year

Application Information: http://www.phrmafoundation.org to download an application and for the specific requirements

3. Name of Program: Awards in Pharmaceutics—Predoctoral Fellowships, Postdoctoral Fellowships, Research Starter Grants

Population: *Predoctoral:* applicants who expect to complete the requirements for the Ph.D. in pharmaceutics in two years or less from the time the fellowship begins; *Postdoctoral:* (1) hold a Ph.D. degree in pharmaceutics from an accredited U.S. university or (2) expect to receive such a degree before activating the fellowship; *Research Starter:* instructor or assistant professor and investigators at the doctoral level with equivalent positions, providing their proposed research is neither directly nor indirectly subsidized to any significant degree by an extramural support mechanism

Program Description:
- Exploring the design and evaluation of contemporary pharmaceutical dosage forms (or drug delivery systems) so that they are safe, effective, and reliable
- Understanding and exploiting the principles underlying drug delivery

Length of Award: *Predoctoral:* two years; *Postdoctoral:* two years; *Research Starter:* two years

Amount Awarded *Predoctoral:* $20,000 per year; *Postdoctoral:* $40,000 per year; *Research Starter:* $30,000 per year

Application Information: http://www.phrmafoundation.org to download an application and for the specific requirements

IPCOR grant
pg. 5